Lecture Notes in Computer Science 4786

Commenced Publication in 1973
Founding and Former Series Editors:
Gerhard Goos, Juris Hartmanis, and Jan van Leeuwen

Editorial Board

Deep Medhi José Marcos Nogueira
Tom Pfeifer S. Felix Wu (Eds.)

IP Operations and Management

7th IEEE International Workshop, IPOM 2007
San José, USA, October 31 – November 2, 2007
Proceedings

Volume Editors

Deep Medhi
University of Missouri–Kansas City
Computer Science & Electrical Engineering Department
550F Flarsheim Hall, 5100 Rockhill Road, Kansas City, MO 64110, USA
E-mail: dmedhi@umkc.edu

José Marcos Nogueira
Federal University of Minas Gerais
Department of Computer Science
Caixa Postal 702, 30123-970 Belo Horizonte, MG, Brazil
E-mail: jmarcos@dcc.ufmg.br

Tom Pfeifer
Waterford Institute of Technology
Telecommunications Software & Systems Group
Carriganore Campus, Waterford, Ireland
E-mail: t.pfeifer@computer.org

S. Felix Wu
University of California–Davis
Department of Computer Science
1 Shields Avenue, 2063 Kemper Hall, Davis, CA 95616, USA
E-mail: wu@cs.ucdavis.edu

Library of Congress Control Number: 2007937345

CR Subject Classification (1998): C.2, D.4.4, D.2, H.3.5, H.4, K.6.4

LNCS Sublibrary: SL 5 – Computer Communication Networks and Telecommunications

ISSN 0302-9743
ISBN-10 3-540-75852-6 Springer Berlin Heidelberg New York
ISBN-13 978-3-540-75852-5 Springer Berlin Heidelberg New York

Springer is a part of Springer Science+Business Media

springer.com

Printed in Germany

Typesetting: Camera-ready by author, data conversion by Scientific Publishing Services, Chennai, India
Printed on acid-free paper SPIN: 12175208 06/3180 5 4 3 2 1 0

Preface

On behalf of the IEEE Communications Society, the Technical Committee on Network Operations and Management (CNOM), the Manweek 2007 Organizing Committee, and the members of the IPOM Technical Program Committee, it is our pleasure to present the proceedings of the Seventh IEEE Workshop on IP Operations and Management (IPOM2007), held as part of Manweek 2007 during October 31-November 2, 2007.

With the widespread deployment of IP-based networks, their operations and management have become increasingly important in terms of understanding emerging technical and scientific problems; lessons from such understanding are particularly important for future Internet design and management.

Like the previous IPOM workshops, IPOM 2007 was co-located with several related events as part of the 3rd International Week on Management of Networks and Services (Manweek 2007). The other events were the 10th IFIP/IEEE International Conference on Management of Multimedia and Mobile Networks and Services (MMNS 2007), the 18th IFIP/IEEE International Workshop on Distributed Systems: Operations and Management (DSOM 2007), the 2nd IEEE International Workshop on Modeling Autonomic Communications Environments (MACE 2007), and the 1st IEEE/IFIP International Workshop on End-to-End Virtualization and Grid Management (EVGM 2007). Co-locating those events provided the opportunity for an exchange of ideas between research communities that work on related topics, allowing participants to forge links and exploit synergies.

This workshop attracted 40 paper submissions through an open call for papers. A rigorous review process was followed with an average of 2.8 reviews per paper. The Technical Program Committee Co-chairs decided to have a strong program with single-track sessions. With this in mind, many papers were discussed regarding both aspects, the recommendations received from the reviewers and their relevance to the theme of the workshop.

These proceedings comprise 16 accepted full papers, thus the acceptance rate was 40%. Authors of accepted papers are from 13 different countries spanning four continents; they were invited to present their work at the conference. In addition, five submissions were chosen for display as posters at the workshop; short papers of 4-page length for these posters have also been included in the proceedings.

The accepted papers present topics around peer-to-peer connectivity and the future Internet, management of Internet security, service management and provisioning, aspects of QoS management and multimedia, as well as management for wireless networks.

We take this opportunity to thank all the members of the IPOM Technical Program Committee, the IPOM Steering Committee, the Manweek 2007 Organizing Committee, and the additional paper reviewers for their hard work that made this workshop possible. In addition, we thank the Springer LNCS team for their support of these proceedings.

Finally, we thank the patrons of Manweek 2007, Cisco Systems, France Telecom R&D, and the Telecommunications Software & Systems Group (TSSG), for their financial and in-kind contributions to this workshop. We thank the IEEE for their continued support and sponsorship of IPOM.

October 2007

Deep Medhi
Jose Marcos Nogueira
Tom Pfeifer
S. Felix Wu

IPOM 2007 Organization

Technical Program Committee Co-chairs

Deep Medhi	University of Missouri-Kansas City, USA
Jose-Marcos Nogueira	Federal University of Minas Gerais (UFMG), Brazil
S. Felix Wu	University of Califonia at Davis, USA

Steering Committee

Prosper Chemouil	OrangeLabs, France
Tom Chen	Southern Methodist University, USA
Petre Dini	Cisco Systems, USA
Andrzej Jajszczyk	AGH University of Science and Technology, Poland
G.-S. Kuo	National Chengchi University (NCCU), Taiwan
Deep Medhi	University of Missouri-Kansas City, USA
Curtis Siller	IEEE Communications Society, USA

Publication Chair

Tom Pfeifer	Waterford Institute of Technology, Ireland

Publicity Chair

Sumit Naiksatam	Cisco Systems, USA

Treasurers

Raouf Boutaba	University of Waterloo, Canada
Brendan Jennings	Waterford Institute of Technology, Ireland

Website and Registration Co-chairs

Edgar Magana	UPC/Cisco Systems, USA
Sven van der Meer	Waterford Institute of Technology, Ireland

Submission Chair

Lisandro Zambenedetti Granville	Federal University of Rio Grande do Sul, Brazil

Manweek 2007 General Co-chairs

Alexander Clemm	Cisco Systems, USA
Silvia Figueira	Santa Clara University, USA
Masum Z. Hasan	Cisco Systems, USA

Manweek 2007 Advisors

Raouf Boutaba	University of Waterloo, Canada
Brendan Jennings	Waterford Institute of Technology, Ireland
Sven van der Meer	Waterford Institute of Technology, Ireland

IPOM 2007 Technical Program Committee

Abdelhakim Hafid	University of Montreal, Canada
Aldri dos Santos	Federal University of Paraná, Brazil
Alexander Clemm	Cisco Systems, USA
Andrzej Jajszczyk	AGH University of Science and Technology, Poland
Baek-Young Choi	University of Missouri-Kansas City, USA
Carlos Westphall	Federal University of Santa Catarina, Brazil
Caterina Scoglio	Kansas State University, USA
David Hutchison	Lancaster University, UK
David Malone	University Maynooth, Ireland
David Maltz	Microsoft Research, USA
Edmundo Madeira	State University of Campinas, Brazil
G.-S. Kuo	National Chengchi University, Taiwan
Gerry Parr	University of Ulster, UK
Hakima Chaouchi	National Institute of Telecommunication, France
Luciano Paschoal Gaspary	Federal University of Rio Grande do Sul, Brazil
Manu Malek	Stevens Institute of Technology, USA
Marcus Brunner	NEC Europe Ltd., Germany
Martin Stiemerling	NEC Europe Ltd., Germany
Masum Hasan	Cisco Systems, USA
Mícheál Ó Foghlú	Waterford Institute of Technology, Ireland
Michal Pioro	Warsaw University of Technology, Poland and Lund University, Sweden
Petre Dini	Cisco Systems, USA
Sasitharan Balasubramaniam	Waterford Institute of Technology, Ireland
Stephan Steglich	TU Berlin / Fraunhofer FOKUS, Germany
Thomas Magedanz	Fraunhofer FOKUS, Germany
Timothy Gonsalves	Indian Institute of Technology Madras, India
Tom Chen	Southern Methodist University, USA
Wolfgang Kellerer	DoCoMo Euro-Labs, Germany
Yacine Ghamri-Doudane	LRSM, Institut d'Informatique d'Entreprise, France

IPOM 2007 Additional Paper Reviewers

Andrzej Bak	Warsaw University of Technology, Poland
Dijiang Huang	Arizona State University, USA
Fabricio Gouveia	Fraunhofer FOKUS, Germany
Fábio Verdi	State University of Campinas, Brazil
Fernando Koch	Universidade Federal de Santa Catarina, Brazil
Haiyang Qian	University of Missouri-Kansas City, USA
Jens Fiedler	Fraunhofer FOKUS, Germany
Jose Carrilho	State University of Campinas, Brazil
Luis Carlos De Bona	Federal University of Paraná, Brazil
Luiz Bittencourt	State University of Campinas, Brazil
Luiz Carlos Albini	Universidade Federal do Paraná, Brazil
Madalena Pereira da Silva	Faculdades Integradas UNIVEST, Brazil
Marcelo Perazolo	IBM Corporation, USA
Plarent Tirana	University of Missouri-Kansas City, USA
Rolf Winter	NEC Europe Ltd., Germany
Shekhar Srivastava	Schema Ltd., USA
Simon Schuetz	NEC Europe Ltd., Germany

Table of Contents

P2P and Future Internet

Internet Security Management

Service Management and Provisioning

QoS Management and Multimedia

Management for Wireless Networks

Short Papers

A Novel Peer-to-Peer Naming Infrastructure for Next Generation Networks

Ramy Farha and Alberto Leon-Garcia

University of Toronto, Toronto, Ontario, Canada
ramy.farha@utoronto.ca, alberto.leongarcia@utoronto.ca

Abstract. One of the major challenges in next generation networks is naming, which allows the different entities on the network to identify, find, and address each other. In this paper, we propose a novel Peer-to-Peer naming infrastructure, which takes into account the expected changes in the next generation networks due to the trends shaping network evolution. The success of the Peer-to-Peer paradigm for applications such as file sharing and instant messaging has lead to research on other areas where such a paradigm could be useful, to provide the scalability, robustness, and flexibility that characterize Peer-to-Peer applications.

1 Introduction

Significant changes are expected to the current Internet architecture, with several emerging trends shaping its future, but with no common understanding as to how this next generation architecture will look like. While clearly a success, the current Internet is not built to play the role that new and emerging services require from it. Traditional research has mainly focused on creating patches or adding functionalities to the existing Internet to resolve the problems faced with each change introduced, but some researchers are advocating a clean-slate design for the next generation Internet [1]. In parallel, the next generation of wireless networks is rapidly emerging, driven by an explosive growth in the number of wireless devices, and by higher data rates made possible through the design of new access network solutions [2]. Therefore, one of the major challenges in next generation networks relates to the way in which naming is performed, in order to allow the different entities to "know" about each other.

The goal of this paper is to propose a novel Peer-to-Peer (P2P) naming infrastructure for next generation networks. The infrastructure takes into account all mobility types (terminal, session, personal, and service) [3], as well as emerging trends such as virtualization [4]. A P2P approach is proposed where distributed Naming Agents (NAs) store the mappings needed to keep the naming infrastructure updated. Such a naming infrastructure is essential for efficient management of next generation networks and systems.

The rest of this paper is structured as follows. In section 2, we summarize some related work. In section 3, we describe the P2P naming infrastructure and present the entities and identifiers needed. In section 4, we explain the different

D. Medhi et al. (Eds.): IPOM 2007, LNCS 4786, pp. 1–12, 2007.

mappings needed to keep the P2P naming infrastructure updated. In section 5, we summarize and conclude this paper.

2 Related Work

The issue of naming has received a significant amount of research since it is a core enabler of any large-scale network. The flexibility, expressiveness, and correctness of the naming infrastructure are key factors that determine its popularity. The purpose of the original Internet is different from that envisioned in the future. Hence, the traditional naming infrastructure, based on the Internet Protocol (IP) addresses and the Domain Name System (DNS) names, has often been criticized, and enhancements have been proposed to address its shortcomings. We now summarize some of the most interesting proposals for naming.

The Host Identity Protocol (HIP) [5] separates the roles of endpoint identifiers and locators of the IP address. Thus, there is a need for a new layer between the transport and the IP layers. HIP introduces Host Identifiers, which are statically globally unique names for naming any system with the IP stack. IP addresses continue to act as locators, whereas Host Identifiers take the role of endpoint identifiers. This separation of endpoint identifiers and locators is a promising idea we use in our P2P naming infrastructure.

The layered naming architecture [6] defines four layers of naming, which are, from top to bottom, the user-level descriptor, the service identifiers, the endpoint identifiers, and the IP addresses. Mappings between the adjacent layers need to be kept. This concept is very close to our proposed P2P naming infrastructure, but fails to accommodate for virtualization of physical resources.

The Ambient Networks approach [7] also adopts a layered approach that consists of four layers, which are, from top to bottom, the applications, the bearer abstraction, the flow abstraction, and the connectivity. This flexible naming architecture is based on dynamic interactions between names, addresses, and identities, with dynamic binding between entities at the different levels preserving the naming architecture. This work is still introductory, but the approach taken seems promising and deserves a more thorough examination.

The Intentional Name Service (INS) [8] scheme allows scalable resource discovery, with several resolvers collaborating as peers to distribute resource information and to resolve queries. The Twine architecture associated with INS achieves scalability by partitioning resource descriptions and hashing unique subsequences of attributes and values from these resource descriptions. INS / Twine is therefore a resource description mechanism (INS) to which a resource discovery procedure is tied (Twine).

In P2P networks, the resulting interconnected set of peers forms an overlay network. P2P topologies can be categorized as unstructured or structured. In structured P2P topologies such as Chord [9], the process of finding a match to a query for a resource has a predictable performance since the overlay topology is tightly controlled and the placement of resources is in precise locations in the overlay. A variant of the original flat P2P topology known as a hierarchical P2P topology has been developed, where Super-Peers are interconnected to form an overlay network. The hierarchical topology recognizes the heterogeneity of peers in terms of communications and processing resources and adaptively elevates peers to the role of Super-Peer. Super-Peers have mainly been presented for unstructured P2P topologies [10], and have rarely been used for structured P2P topologies [11]. We will use Super-Peers for a structured P2P topology used in our P2P naming infrastructure.

The use of P2P concepts is starting to expand to areas related to naming and mobility management. Mobility management using P2P networks has first been proposed in [12] by organizing home agents into P2P networks for horizontal mobility, and has then been extended in [13] for vertical mobility. Another work in [14] addresses P2P for mobility management. The difference with our approach is the use of DNS as a gateway between multiple access networks. Resource discovery using P2P networks has been proposed in [15] by using the P2P paradigm to facilitate resource discovery in Grid systems. The difference with our approach is that P2P concepts are only used for resource discovery, while we propose to use the P2P paradigm for both mobility management and resource / service discovery.

3 Naming Infrastructure Design

We can identify the major entities and the identifier spaces that constitute the next generation networks as follows:

- Customers / Service Providers: We define Service Providers as entities that offer services to other entities, and Customers as entities that receive services from Service Providers based on a contractual relationship, during which Service Providers put their resources at the disposition of Customers. Customers receive services from Service Providers at the Service Access Point, which is needed to separate between the Customer domain and the Service Provider domain. We divide Customers into two main types, depending on the services involved: "Normal" Customers, which are mainly individuals, that activate individual service instances, and "Super" Customers, which are mainly enterprises (or other Service Providers), that activate aggregate service instances. Since Customers can sometimes become Service Providers themselves and can offer services that they had previously bought from other Service Providers, they are referred to as the same entity in the proposed naming infrastructure, with different roles depending on the given service (Customers or Service Providers).
- Physical Resources: We define Physical Resources as the underlying infrastructure available to Service Providers wishing to deliver services to Customers. It also includes the devices that Customers use to access the next generation Internet, whether they are using a wireline or a wireless medium. Therefore, Physical Resources can be either fixed or mobile. In both cases, Physical Resources need a bootstrap, or an access point, to attach to the next generation network infrastructure.

- Virtual Resources / Services: We define a service as the engagement of resources for a period of time, according to a contractual relationship between Customers and Service Providers to perform a function, and a resource as the physical and non-physical (logical) component used to construct those services. The component services used as building blocks to other services are seen as resources to these services. A service is therefore associated with a service template which is a workflow of component services. Note that both services and resources can be referred to as "virtual" due to the emergence of the virtualization concept [4]. The players involved in the delivery of each service are therefore the Customer and its Service Provider. We distinguish between two types of services: Basic services that cannot be broken down anymore into other services, in other words, atomic, and Composite services that are composed of Basic services as well as of other Composite services. Basic services are of two kinds: Allocation, involving resource amounts being allocated to customers, or Configuration, involving resource configurations being done according to customer needs. Basic services have a given "type" field that is used to identify Basic services performing the same operation. There are two possible topologies for Virtual Resources: Localized, where resources are composed of entities which are geographically co-located, or Distributed, where resources are composed of entities which are geographically distributed. We also differentiate between two possible categories of services: Internal services, which are used internally by Service Providers as components to other services, and External services (or Applications), which are offered to Customers by Service Providers. Note that different combinations of types, topologies, and categories are possible. For instance, a (Basic, Localized, Application) combination means that the service is atomic, located in a single geographic area, and offered to customers (for instance a web page hosted on a server for browsing by customers). Since virtual resources can sometimes be directly offered to Customers as services, they are both referred to as the same entity in the proposed naming infrastructure.
- Service Instances: We define Service Instances as an entity that consists of the allocation / configuration of Virtual Resources / Services to Customers by a Service Provider according to an agreement. These Service Instances can be one of two types depending on the type of customers that activated them: Normal Customers activate "Individual" service instances, and Super Customers activate "Aggregate" service instances. In addition, the service instances can be either Quality of Service (QoS) or Best-Effort (BE) depending on the guarantees given to customers activating them.

Based on the entities forming the new naming infrastructure, we define the identifier spaces which provide the identifiers needed to refer to those entities. Given the four entities (Customers / Service Providers, Physical Resources, Virtual Resources / Services, and Service Instances) constituting the next generation networks, we create four corresponding identifier spaces, as follows:

- Identifier Space 1 (IdSpace1): Uniquely identifies Customers / Service Providers.
- Identifier Space 2 (IdSpace2): Uniquely identifies Physical Resources in the underlying infrastructure.
- Identifier Space 3 (IdSpace3): Uniquely identifies Virtual Resources / Services, whether they are Localized or Distributed, Basic (Allocation or Configuration) or Composite, Internal or External.
- Identifier Space 4 (IdSpace4): Uniquely identifies Service Instances, whether Individual or Aggregate, BE or QoS.

Next, we will describe these four identifier spaces, summarized in Table 1, in more details, and detail the identifiers that these identifier spaces consist of.

Table 1. Identifier Spaces

Identifier Space	**Entities**
IdSpace1	Customers / Service Providers
IdSpace2	Physical Resources
IdSpace3	Virtual Resources / Services
IdSpace4	Service Instances

The first Identifier Space, **IdSpace1**, needs to uniquely identify Customers and Service Providers. However, Customers and Service Providers are increasingly mobile, and use different physical devices to access the next generation Internet. A mapping therefore needs to be kept between unique, permanent, and global scope identifiers, referred to as Customers / Service Providers Permanent Identifiers (*perm_csp_id*), and unique, temporary, and global scope identifiers, referred to as Customers / Service Providers Temporary Entry Points (*temp_csp_id*). The permanent identifiers could be something like a SIP URI, an email address, a host name, or anything that refers to a Customer or a Service Provider in an unequivocal way, regardless of the physical device and location used to access the next generation Internet, whereas the temporary identifiers depend on the layer 3 protocol such as IP or any new protocol that Customers / Service Providers devices use to access the Internet. If such layer 3 protocol is IP, the temporary identifiers are the IP addresses.

The second Identifier Space, **IdSpace2**, needs to uniquely identify Physical Resources. Since Physical Resources can be moved in the Internet, as well as being added and removed dynamically, within or across different access network technologies such an identifier space therefore needs to keep a mapping between three identifiers: unique, permanent, and global scope identifiers referred to as Physical Resources Global Identifiers (*global_pr_id*), unique, permanent, and access network technology scope identifiers, referred to as Physical Resources Permanent Identifiers (*perm_pr_id*), and unique, temporary, and access network technology scope identifiers, referred to as Physical Resources Temporary Attachment Points (*temp_pr_id*). The global identifiers could be something like a Physical Resource's serial number, or anything that refers to a Physical Resource in an unequivocal way, regardless of the Physical Resource's location, whereas the permanent and temporary identifiers depend on the access network technology as well as on the layer 2 (MAC) or layer 3 protocols such as IP or any new protocol that Physical Resources use to attach to the Internet. If such layer 3 protocol is IP, the temporary identifiers are the IP addresses. If the Physical Resources only move within an access network technology (horizontal mobility), then they only need to keep a mapping between two out of the three identifiers of **IdSpace2**: *perm_pr_id*, and *temp_pr_id*.

The third Identifier Space, **IdSpace3**, needs to uniquely identify Virtual Resources / Services. Each service has a service template associated with it. This template is represented by a workflow of resource allocations and configurations. This identifier space is also made possible by the emergence of virtualization, which allow Virtual Resources / Services to be tied to Physical Resources according to different topologies. Several Virtual Resources / Services characteristics are possible, but in all cases, the Virtual Resource / Service consists of one or more Virtual Resources hosted on Physical Resources, identified by unique, permanent, and global scope identifiers (*vs_id*). When Virtual Resources / Services are basic, their type is uniquely and globally identified using a global scope type identifier (*vs_type_id*).

The fourth Identifier Space, **IdSpace4**, needs to uniquely identify Service Instances. This identifier space is made possible by the emergence of virtualization, which allows each Service Instance consisting of an allocation of portions of virtual resources from the underlying physical resources and of a given configuration of the underlying physical resources, to be uniquely identified. As mentioned previously, customers are divided into two types: Normal, activating Individual Service Instances, or Super, activating Aggregate Service Instances. Both are identified using a Service Instance Identifier *si_id*. Therefore, this identifier space needs a two-layer naming structure, with a parent-child relation between aggregate service instances and individual service instances constituting them.

In addition to the aforementioned identifiers, we also define a **Technology Identifier** *tech_id* to identify each possible access network technology. The challenge to achieve both naming, as well as horizontal and vertical mobility, becomes that of managing the mappings between all these different identifiers, in order to allow all entities to be found anywhere and at anytime. Naming and mobility management should also allow, in addition to terminal mobility, other types of mobility, such as session, personal, and service mobility [3].

An important observation to make is that the temporary identifiers of Customers / Service Providers in **IdSpace1** and Physical Resources in **IdSpace2** are similar, corresponding to their current location and the current Physical Resource (or device) that this Customer / Service Provider entity is using. The identifiers needed for the proposed naming infrastructure are summarized in Table 2. The attributes of the different identifiers are summarized in Fig. 1.

Table 2. Identifiers Needed

Identifiers	Description
perm_csp_id	Permanent Identifier of Customer / Service Provider
temp_csp_id	Temporary Identifier of Customer / Service Provider
temp_pr_id	Temporary Identifier of Physical Resource
perm_pr_id	Permanent Identifier of Physical Resource
global_pr_id	Global Identifier of Physical Resource
vs_id	Identifier of Virtual Service offered
vs_type_id	Virtual Resource Type Identifier
si_id	Identifier of Service Instance activated
tech_id	Access Network Technology Identifier

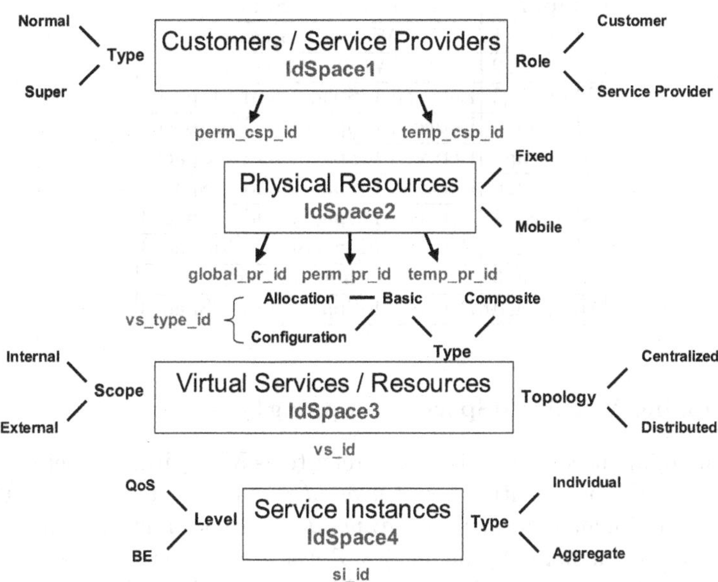

Fig. 1. Summary of identifiers and their attributes

4 Mappings Between Identifier Spaces

Several mappings need to be kept between the aforementioned identifier spaces in the P2P naming infrastructure. The use of word "mapping" in this context is aimed at conveying the need to keep a relation between the different identifier spaces and entities. We identify several possible mappings, shown in Table 3. These mappings are either within an identifier space, or between different identifier spaces. Keeping the mappings updated at all times allows the management system to make informed decisions on the underlying next generation network infrastructure. Mappings are kept at trusted machines that we refer to as

Naming Agents (NAs) for the rest of this paper, which are either within a Service Provider's domain, or in different Service Provider's domains. Fig. 2 shows how the NAs use a P2P substrate to keep the mappings needed updated at all times. The Primary and Secondary fields in the mapping refer to the globally unique identifier used to locate the mapping (primary), and the content of the mapping (secondary). In a realistic setting, some mappings are more likely to be distributed than others. The P2P substrate is built according to **h-Chord**, a Super-Peer variant of the aforementioned Chord topology, which we had presented in [14]. Super-Peers (NAs) store the mappings using hash functions such as SHA-1 [16] in a distributed P2P ring.

Table 3. Mappings between Identifier Spaces

Mapping	**Identifier Spaces**
Mapping1	Within IdSpace2
Mapping2	Within IdSpace2
Mapping3	Between IdSpace1 and IdSpace2
Mapping4	Between IdSpace1 and IdSpace3
Mapping5	Between IdSpace1 and IdSpace4
Mapping6	Between IdSpace3 and IdSpace4
Mapping7	Between IdSpace1 and IdSpace3
Mapping8	Between IdSpace1 and IdSpace3
Mapping9	Within IdSpace3
Mapping10	Between IdSpace2 and IdSpace3

4.1 Mapping Within IdSpace1 (Mapping1)

The first mapping needed, which we will refer to as **Mapping1**, is between Physical Resources Global Identifiers *global_pr_id* (Primary), and a list of Physical Resources Permanent Identifiers, *perm_pr_id*, corresponding to each supported access network technology (Secondary). If Physical Resources are fixed, this mapping is static. This mapping is within IdSpace2, and its update frequency depends on the rate at which mobile Physical Resources move and change their entry point across different access network technologies (vertical mobility).

4.2 Mapping Within IdSpace2 (Mapping2)

The second mapping needed, which we will refer to as **Mapping2**, is between Physical Resources Permanent Identifiers *perm_pr_id* (Primary), and Physical Resources Temporary Attachment Points Identifiers *temp_pr_id* (Secondary). If Physical Resources are fixed, this mapping is static. This mapping is within IdSpace2, and its update frequency depends on the rate at which mobile Physical Resources move and change their entry point within an access network technology (horizontal mobility).

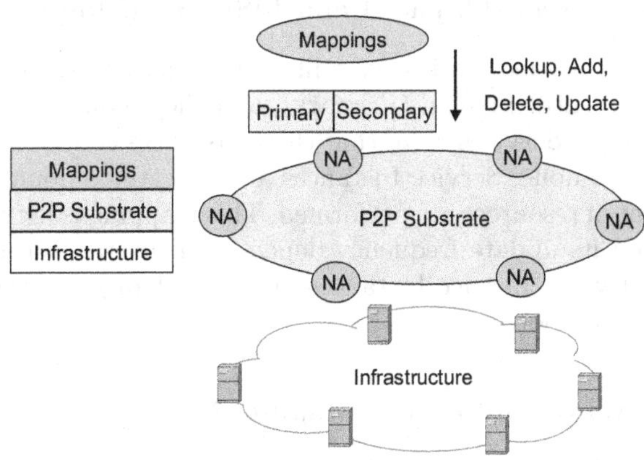

Fig. 2. P2P Substrate in the Naming Infrastructure

4.3 Mapping Between IdSpace1 and IdSpace2 (Mapping3)

The third mapping needed, which we will refer to as **Mapping3**, is between Customers / Service Providers Permanent Identifiers *perm_csp_id* (Primary), and a list of Physical Resources Global Identifiers, *global_pr_id*, owned / used by Customers / Service Providers (Secondary). This mapping is between IdSpace1 and IdSpace2, and its update frequency depends on the rate at which Customers change the Physical Resources, or devices, that they use to receive or offer services on the network, or the rate at which Service Providers add or remove Physical Resources to their network.

4.4 Mapping Between IdSpace1 and IdSpace3 (Mapping4)

The fourth mapping needed, which we will refer to as **Mapping4**, is between Customers Permanent Identifiers *perm_csp_id* (Primary), and a list of external Virtual Resources / Services, *vs_id*, bought by these customers (Secondary). This mapping is between IdSpace1 and IdSpace3, and its update frequency depends on the rate at which Customers buy Virtual Resources / Services.

4.5 Mapping Between IdSpace1 and IdSpace4 (Mapping5)

The fifth mapping needed, which we will refer to as **Mapping5**, is between Customers Permanent Identifiers *perm_csp_id* (Primary), and a list of Service Instances *si_id* activated by those customers (Secondary). Customers, after buying services which are kept by Mappping4, activate Service Instances. This mapping is between IdSpace1 and IdSpace4, and its update frequency depends on the lifetime of the activated Service Instances.

4.6 Mapping Between IdSpace3 and IdSpace4 (Mapping6)

The sixth mapping needed, which we will refer to as **Mapping6**, is between Service Instances Identifiers *si_id* (Primary), and the amounts / configurations of Virtual Resources / Services *vs_id* that these instances consist of (Secondary). Once provisioning is done, Service Instances are allocated amounts of resources and configurations of resources are performed. This mapping is between IdSpace3 and IdSpace4, and its update frequency depends on several factors such as the lifetime of the activated Service Instances, and the changes to the Virtual Resources / Services components.

4.7 Mapping Within IdSpace1 (Mapping7)

The seventh mapping needed, which we will refer to as **Mapping7**, is between Basic Virtual Service Types Identifiers *vs_type_id* (Primary), and Service Providers Permanent Identifiers *perm_csp_id* offering this Basic Virtual Service type (Secondary). This mapping is between IdSpace1 and IdSpace3, and its update frequency depends on the rate at which Service Providers create / buy new Basic Services and offer them to other customers / Service Providers.

4.8 Mapping Between IdSpace1 and IdSpace3 (Mapping8)

The eighth mapping needed, which we will refer to as **Mapping8**, is between Virtual Resources / Services *vs_id* (Primary), and Service Providers Permanent Identifiers *perm_csp_id* owning these Virtual Resources / Services (Secondary). This mapping is between IdSpace1 and IdSpace3, and its update frequency depends on the rate at which Service Providers create / buy new Virtual Resources / Services to offer to their Customers or the rate at which Customers buy a service and then become Service Providers for that service.

4.9 Mapping Within IdSpace3 (Mapping9)

The ninth mapping needed, which we will refer to as **Mapping9**, is between Virtual Resources / Services *vs_id* (Primary), which are External, hence ready to be offered to customers, and a list of the component Virtual Resources / Services *vs_id* composing this Virtual Resource / Service, which could follow any combination for the topologies (Localized or Distributed), the types (Basic or Composite), and the categories (Internal or Application) (Secondary). This mapping is within IdSpace3, and its update frequency depends on changes to the components of the Virtual Resources / Service.

4.10 Mapping Between IdSpace2 and IdSpace3 (Mapping10)

The tenth mapping needed, which we will refer to as **Mapping10**, is between Virtual Resources / Services *vs_id* which are hosted on physical resources

(Primary), and the hosting Physical Resources Global Identifiers *global_pr_id* (Secondary). This mapping is therefore between IdSpace2 and IdSpace3, and its update frequency depends on the rate of Physical Resources additions, removals, or failures in the network.

The customer, using a physical device, which could be fixed or mobile, activates a service it had bought from a Service Provider, creating a service instance. The service operates using component services on physical resources. The way the aforementioned mappings interact is shown in Fig. 3. This completes our detailed description of a new P2P naming infrastructure for next generation networks based on emerging trends. Due to space limitations, we omit the results of some simulations performed to assess performance capabilities of this infrastructure, which have shown improvements in load balancing capabilities at the expense of P2P lookup costs, when compared to centralized approaches.

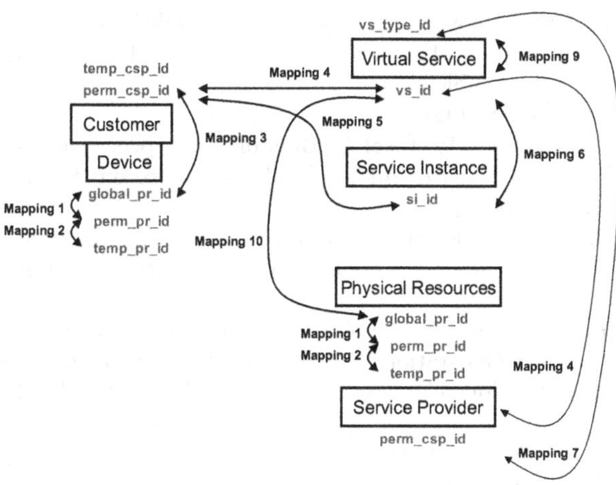

Fig. 3. Relation between the different mappings

5 Conclusion

In this paper, we have presented a new Peer-to-Peer naming infrastructure for next generation networks, allowing both horizontal and vertical mobility management. We presented the entities and identifier spaces needed, explained the different mappings required between the four proposed identifier spaces, and illustrated how those mappings can be kept updated using Peer-to-Peer mechanisms, in order to benefit from the scale, robustness, and performance that such distribution offers. The potential of such a Peer-to-Peer naming architecture deserves to be further explored, since the benefits it will have on the management systems of next generation networks are significant.

References

[1] National Science Foundation. The GENI Initiative, http://www.geni.net
[2] Akyildiz, I.F., et al.: A survey of mobility management in next-generation all-IP-based wireless systems. IEEE Wireless Communications, 16–28 (2004)
[3] Schulzrinne, H., Widlund, E.: Application-Layer Mobility Using SIP. ACM SIGMOBILE Mobile Computing and Communications Review, 47–57 (2000)
[4] Anderson, T., et al.: Overcoming the Internet Impasse through Virtualization. IEEE Computer, 34–41 (2005)
[5] Moskowitz, R., Nikander, P.: Host Identity Protocol (HIP) Architecture. RFC 4423 (2006)
[6] Balakrishnan, H., et al.: A layered naming architecture for the Internet. In: Proceedings of the conference on applications, technologies, architectures, and protocols for computer communications, pp. 343–352 (2004)
[7] Ahlgren, B., et al.: Names, Addresses and Identities in Ambient Networks. Dynamic Interconnection of Networks Workshop (2005)
[8] Balazinska, M., et al.: INS/Twine A Scalable Peer-to-Peer Architecture for Intentional Resource Discovery. In: Mattern, F., Naghshineh, M. (eds.) Pervasive Computing. LNCS, vol. 2414, pp. 149–153. Springer, Heidelberg (2002)
[9] Stoica, I., et al.: A scalable peer-to-peer lookup service for internet applications. MIT, Tech. Rep. TR-819 (2001)
[10] Montresor, A.: A Robust Protocol for Building Superpeer Overlay Topologies. In: Proceedings of the 4th IEEE International Conference on Peer-to-Peer Computing, pp. 202–209 (2004)
[11] Garces-Erice, L., et al.: Hierarchical P2P Systems. In: Proceedings of ACM/IFIP International Conference on Parallel and Distributed Computing, 1230–1239 (2003)
[12] Farha, R., et al.: Peer-to- Peer Mobility Management for all-IP Networks. In: Proceedings of the Next-Generation Mobile Networks Symposium at the IEEE International Communications Conference (ICC), pp. 1946–1952. IEEE, Los Alamitos (2006)
[13] Farha, R., et al.: Peer-to- Peer Vertical Mobility Management. In: Proceedings of the Network Services and Operations Symposium at the IEEE International Communications Conference (ICC) (to appear, 2007)
[14] Lo, S.: Mobility Management Using P2P Techniques in Wireless Networks. Journal of Information Science and Engineering, 421–439 (2007)
[15] Mastroiannia, C., et al.: A super-peer model for resource discovery services in large-scale Grids. Elsevier Journal on Future Generation Computer Systems, 1235–1248 (2005)
[16] National Technical Information Service, U.S. Dept. Commerce / NIST: Secure Hash Standard, FIPS 180-1 (1995)

Description of a Naming Architecture Managing Cryptographic Identifiers

Daniel Migault[1] and Marcus Brunner[2]

[1] France Telecom R&D
mglt.biz@gmail.com
[2] NEC Europe Ltd.
brunner@netlab.nec.de

Abstract. The necessity to split the endpoint identity and locator has been understood since sometime both from routing and security perspective. Today endpoints are identified by IP address that is location dependent and attributed by ISPs, whereas the identity neither depends on location nor on ISP. So splitting the routing and identification space is expected to make network operation such as mobility, multihoming and traffic engineering transparent for the end user. While in the operator side the use of a single space for routing and identification brings scaling issues. The operators will benefit from the split by decreased routing table size.

Within IETF/IRTF solutions are being developed to separate the IP layer into Endpoint Identifier (EID) space and routing locator (RLOC) space in the form of Locator/ID Separation Protocol (LISP). In LISP the Identifier (ID) has the format of a IPv4 or IPv6 address. This architecture provides ID to locator resolution so that the packets can be routed through the Internet.

This paper proposes a solution that considers an Endpoint Identifier (EID) as the combination of a domain name and a cryptographic Identifier (cryptoID). Such EIDs are hosted in a mixed DNS/Distributed Hash Table (DHT) architecture. Resolution involves a DNS and a DHT resolution. We show how the use of DNSSEC enhances the routing algorithm of the DHT resolution, and present advantages a such an architecture in term of deployment and future use of the Internet.

1 Introduction

Internet today uses IP addresses for both identifying a node and routing packets. With the requirements such as mobility, and multihoming, and advent of wireless Internet a node can have multiple IP addresses on different network interfaces, such as Wi-Fi or WiMAX, and it is likely that its IP address changes, or at least the Care of Address (CoA), while moving from one network to one network. Such change in IP address will lead to discontinuation of an on-going session. Thus it is necessary to consider endpoint to endpoint connection independent of the IP address. This leads to consideration of splitting the IP layer into two spaces: the Endpoint IDentifier space (EID), and the Routing LOCator space (RLOC).

D. Medhi et al. (Eds.): IPOM 2007, LNCS 4786, pp. 13–24, 2007.

Protocols like SHIM6 [1] or Host Identity Protocol (HIP) [2] are already considering the use of EIDs for mobility and multihoming purposes. SHIM6 is initiating a connection with a routing locator, in a traditional way. When the initiating locator is not anymore routable, then it is considered as an EID and is bound to other routable locators. HIP is considering a new naming space of EIDs with security properties. Host Identity Tag (HIT) are hash of a public key and so called cryptographic identifiers (CryptoID). CryptoID are very useful since they can be self generated and provide proof-of-ownership even on adhoc scenario. Nevertheless, to proceed to a HIP connection an EID-to-locator resolution is required, and no naming system really fits this resolution.

EID-to-locator resolution with EIDs as cryptoID is really what is missing in the current Internet architecture.

As opposed to DNS naming space that relies on a hieratical structure, cryptoIDs rely on a flat structure. A common way to deal with flat data is to use Distributed Hashed Table (DHT) data bases.

We chose in this paper to consider the chord [3] DHT structure because all data are distributed and ordered on a circle. We do not claim that chord is the only DHT database to be used. The chord data base enables to make the binding between HIT and data associated to this HIT.

Using a global DHT chord database for HITs implies consensus of all Internet authorities/actors to delegate the Naming function to a single entity. This is hardly believable, at least in the short term. Furthermore it would also imply the coexistence of two different and independent Naming architectures : DNS and DHT.

The hierarchical Naming structure of the DNS can be used to split the administrative part of the EID-to-locator resolution. On the other hand, using DNS means a single entry for Naming resolution.

Our paper is not presenting a proposition on how to use the DNS system over DHT. It describes a new identifier format and a naming solution that can use the traditional DNS architecture as an entry point. For that purpose we present an architecture that combines DNS and DHT. The naming resolution requires two different protocols DNS. In our example the DHT related protocol is chord. We point out how DNS can provides enough information to enable and enhance the chord resolution by using DNSSEC ([4], [5], [6]). Although DNSSEC was originally designed for security purposes, in this paper DNSSEC properties are used to indicate to resolvers the chord node the naming request has to be sent to. We also show how this property is improving the chord routing algorithm. Our work did not intend to look on how to enhance DHT routing algorithm for naming service [7], [8] nor to compare [9] or migrate DNS to DHT databases.

On the other hand as pointed out by DNS over DHT works, the use of DHT fits by nature the new requirements of cryptoIDs. DHT offers to naming system scalability [10], robustness to DDoS [11] and also reduce the administrative burden for delete update operations [11].

This paper provide system requirements in section 2, then positioned the paper with related works in section 3. The EID format is discussed in section 4.

The mixed architecture is presented in section 6, after a brief overview of DNS and chord architectures in section 5. The conclusion takes in consideration future works.

2 System Requirements

Section 4 is providing more details on the format and motivation of using cryptographic identifier, but basically cryptographic identifiers are very practical since they can be self generated, and carry some security properties.

Protocols like HIP are considering the cryptoIDs but provide no way to bind the crytpoID to locators like IP addresses. As opposed to Fully Qualified Domain Names (FQDN), or IP addresses, cryptoIDs are not assigned, which means there are no administrative authorities, like ISP for example, that can be responsible for managing your cryptoID.

CryptoID are by nature random number, and there are no ways to introduce any hierarchical structure such as in DNS. DHT databases seem to fit the non-hierarchical nature of cryptoIDs, but would on the other hand require a global unique database for all those cryptoIDs. It is hardly believable that ISP, governments and other Internet authorities would assign the naming function to one single entity, at least in a near future. On the other hand using a cryptoID dedicated database imply that there are two parallel naming space and that resolvers need to differentiate the naming resolution for FQDN and cryptoIDs.

So system requirements are :

- Providing a EID format and an architecture that enables to split EIDs naming space between different organizations.
- Providing a EID format and an architecture that enable migration to EIDs as only formed of cryptoIDs.
- The naming architecture needs a single entry point compatible with existing FQDN.
- The naming system should enable delegation of EIDs by different independent administrative entities.
- The naming architecture should have to enable easy administration, robust to DDoS, enable dynamic update operation.
- The naming architecture does not have strong requirements on scalability.
- The naming architecture should provide fast resolution mechanisms.

Our architecture is providing an EID format constituted of a domain name and a cryptoID. EIDs constituted by cryptoID only (like HIT) can be considered as directly appended to the root domain (.). We can use DNS and root domain as the single naming entry point for resolution, and uses DNS delegation to delegate cryptoID management to different entities. Those entities are called communities. Each community represented by a domain name manages the cryptoIDs with DHT databases, then improving dynamic dynamic update operations. DNSSEC properties enable the DNS to indicate the nodes in the DHT database that owns the data (i.e. the locators), thus enhancing naming resolution.

3 Related Works and Position of Our Work

One can split works related to this paper into two parts, the work considering a routing architecture based on EIDs and the works related to DNS and DHT.

IRTF/IETF has designed protocols like HIP and SHIM6 using EIDs for mobility, multihoming and security purposes. On the other hand they are also investigating for a global network architecture that considers the creation of the EID and RLOC space. In addition to the mobility and multioming purpose, their expectation is to reduce the routing table size in the "default-free zone" (DFZ) and to enable traffic engineering (TE).

The separation generates a new protocol "Locator/ID Separation Protocol" (LISP) [12]. A user sends an IP packet with EIDs instead of IP addresse to an ITR. After an EID to RLOC resolution the packet is tunneled to the appropriated ETR, and finally to the destination node. The LISP protocol is considering multiple version of it. Non routable EIDs are considered in LISP2 and LISP3. LISP2 considerers the EID to RLOC resolution done though the DNS, whereas LISP3 considers other databases defined in [13], [14] and [15].

LISP works differs on three points from what is presented in this paper.

- This paper is looking for the benefits of using another type of format, and thus uses a combination of domain name and the hash of a public key. This combination is believed to balance security purposes and readability by end user as well. LISP is considering a single format of EID which is similar to the one of an IP address.
- This paper proposes an architecture that requires very few changes in the network architecture. The routing core network is not impacted and resolution is performed by the end users. So only the Naming architecture is impacted and needs additional databases. The LISP architecture is considering deployment of ITR/ETR devices, as well as a new database system that enables the EID-to-locator binding.
- This paper is considering an functional naming architecture with a single naming entry point. LISP is considering EIDs-to-locator resolution independent of FQDN-to-locators resolution.

Many works, simulation and experiment have been made with DNS and DHT. Their main purpose is to look how the set of DNS databases can be replaced by a peer-to-peer database and then solve a few issues encountered by the DNS. Peer-to-peer databases are expected to enhance scalability of the DNS, by spreading not the whole zone file but data among thousands of nodes [10]. The nodes are self-organizing, can leave and join the database, which ease the administrative burden of DNS administrators [11]. On the contrary to DNS databases, DHT data bases are distributed and does not have single entry point. Data are randomly spread among nodes which enhance robustness against DDoS attacks [11]. On the other hand DNS over a chord database does not provide improvement over the traditional DNS architecture in term of performance [9]. In fact DHT has heavy routing algorithm to find the node that hosts the requested data. Mechanisms such as Beehive [7] improves DHT performances with proactive replication of data on nodes. The DNS implementation of this mechanism is called CoDoNS [8].

DNS/DHT works mainly concern how to make the DNS service gaining DHT advantages, and in return how to improve DHT routing algorithm so that DHT can enhance the DNS high availability. Motivations of those works did not concern the use of DHT database to manage flat data such as cryptoIDs.

Some other works like [16] provides mixed DNS and DHT architecture for managing massive flat data like RFIDs or serial number. This architecture is pretty close to the one described in this paper to the exception that the naming architecture is a specific naming architecture for private application, whereas this paper is considering the global Internet naming architecture. Furthermore, the DNS and DHT parts of the architecture are independent, and the DNS part does not enhance the DHT resolution. In our case the DNS part provides delegation information of the more adequate node the DHT request should be sent to, enhancing the DHT routing algorithm. So finally :

- This paper is looking on DHT database to manage flat data such as cryptoIDs. Presented woks above were looking on DHT to enhance the already existing DNS service.
- This paper is presenting a global Naming architecture, and the naming resolution is performed by clients using DNS and DHT resolution for the global naming resolution. It does not require any translators, although migration could require such translators.
- This paper is not combining two independent way of managing data. DNS is enhancing the DHT resolution by providing the home node of the DHT database thanks to a DNSSEC mechanism.
- This work is considering DHT database managed by specific entities, with dedicated servers. There is no requirement for self-organization or scalability properties as DHT databases usually require.

4 Identifiers

4.1 Human Readable Versus Binary Identifiers

Nodes are commonly designated with human readable names, such as `mailserver1`, useful for human beings to identify nodes through user interface or for management purposes. Nevertheless, theses names have many drawbacks. They must be relatively small, be composed of only a few possible combinations of numbers and letters, which requires, in the end, a centralized registered center to avoid duplication.

Binary identifiers have no restrictions due to meanings, and can be made of any sequence of bits. This flexibility is very suitable for programs or automatic configuration. At last they can carry different kind of information, such as routing information, locator, or cryptographic characteristic. In the latter case, they are called cryptographic identifier or cryptoID.

4.2 Cryptographic Identifiers

CryptoID is a public key, or its hash. Unlike routing information, there is no inconvenient that security information is included into the identifier since the information is related to the identity of the node and is location-independent. Including security parameters into the identity is also the best way to have "built in" security. Crypto-IDs can perform authentication, integrity check, confidentiality and proof-of-ownership operation without any heavy security infrastructures. This brings us the concept of "opportunistic security". Communication between two nodes can not be hijacked by a third node, and authentication by other application can also be based on the crypto-ID. Despite cryptoIDs fit very nicely identity purposes they are based on a flat structure.

5 Naming Space

5.1 DNS Architecture

The DNS architecture presented in Figure 1is a hierarchical model. "Root servers" are on the top and domains below are called Top Level Domains (TLD). A DNS request is first sent to root DNS servers and then gradually sent down to TLDs and domains below, until the path is completed to get the proper answer.

This hierarchical structure uses domain names that are human readable names, and bind them to IP addresses. It is not designed for flat structure naming space like crypto-IDs, and so can not provide a structure for a cryptoID-to-locator resolution.

5.2 Chord Architecture

The chord architecture is a Distributed Hash Table (DHT) model designed to deal with flat model structure data, like the crypto-ID. The database can be used for resolution between a binary identifier and locators.

All data are stored in a global shared database and routing algorithms are different then in DNS. Any node owns routing information, but does not have a complete description of the architecture. Thus to find the proper node, one might send a request to a few number of nodes which slows down resolutions. On the other hand, the wide number of nodes might also improve dynamic properties of the chord database over the DNS.

In Figure 2, data and nodes are indexed by numbers ordered on a circle defining the chord name space. In that paper binary identifiers (cryptoID)with fixed size hash are the index or the chord ID. Nodes' IDs are represented by green plots. To find out on which node the data corresponding to a specific ID is hosted, chord compares the ID of the Data and the ID of the node. The Data is hosted on the node whose ID is equal or clockwise next to the Data ID in the

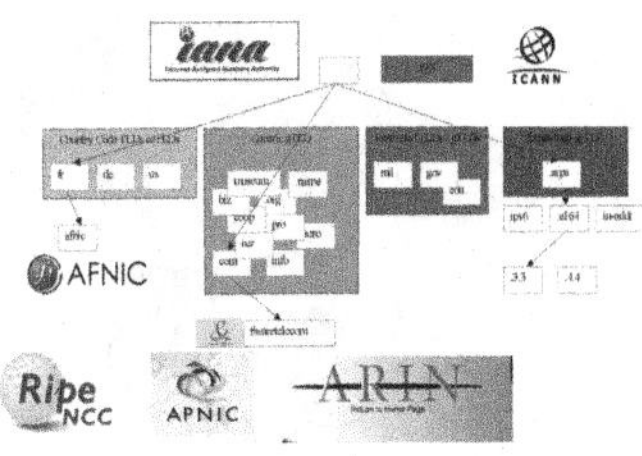

Fig. 1. DNS Naming space

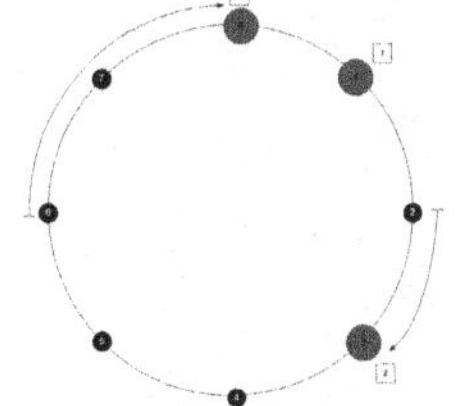

Fig. 2. CHORD Naming space

circle. In our example the next node ID of data *2* on the circle is *3* which means that the request for data *2* has to be sent to node ID *3*.

6 Proposed Mixed DNS/CHORD Architecture

The purpose of the proposed architecture is to associate to one domain name a pool of cryptoIDs. The DNS architecture to provide human readability and enables entities to administrate their own set of identifiers. The chord part of the architecture provide means for cryptoID-to-locator resolution.

Let's consider that a university or a company (and thus a community) provides to each of its members a cryptoID. Our purpose is not to create a completely independent new naming space, but to take advantage of the already existing DNS structure. The key advantages of using the DNS is to associate FQDN to cryptoIDs for readability, to have a single entry point of the naming system, and to provide flexibility for each domain name to administrate its chord database.

On the user point of view, one can assume the user has only one identifier which is its public key. This user can work at "*company.com*" and at the same time be a student at "*university.edu*". The domain name is only a way to attach the cryptoID to a common Naming structure, and for administrative purpose.

Opportunistic security relies mainly on the crypto-ID part, and doesn't rely on this structure, and domain names are only a way to host the key. So one user with a single key can be hosted under different domain names. The domain name can also act as a third party and signed all the data associated to the crypto-ID. DNSSEC can establish a chain of trust from the root servers to the final data. In the example we consider the cryptoID as the hash of a public key, symbolized by "[crypto-ID]". The example describes in Figure 3 is company: "*company.com*" that distributes a cryptoID to each of its employees. They are placed under the domain name "*company.com*". The proposed DNS/chord architecture presented in Figure 3 and Figure 4enables the resolution of *[crypto-ID].compagny.com.* In the example, the crypto-ID value is *2*.

Basically the resolution is proceeded in two steps.

- Traditional DNS resolution of *2.chord.compagny.com.*
- Chord resolution of *2*.

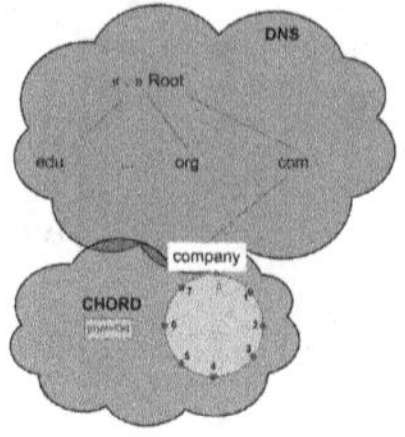

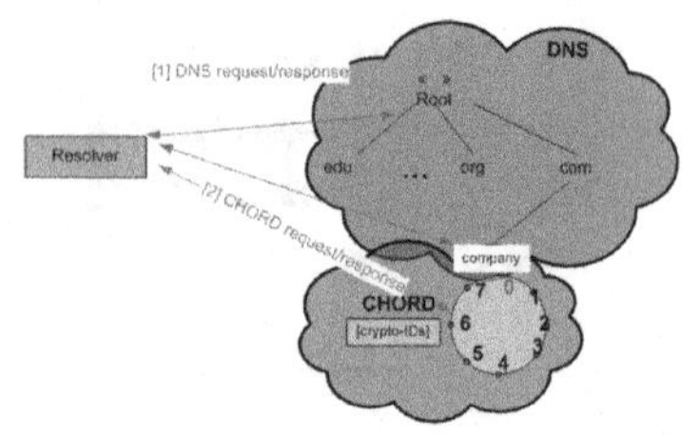

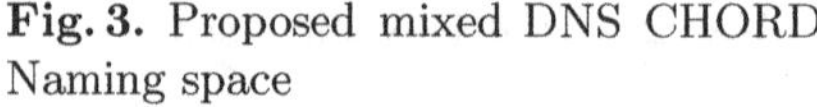
Fig. 3. Proposed mixed DNS CHORD Naming space

Fig. 4. Mixed DNS CHORD Naming resolution

6.1 Implementation and Evaluation of the Mixed DNS/Chord Architecture

The purpose of this section is to propose how traditional DNS servers can be used to :

- Provide the chord node ID that is hosting the requested data.
- Provide the location of the chord node ID.

To represent the chord nodes hosting data one introduces the new DNS RRset type `CHORDNODE`. To send the chord request to the node, one needs the location of the node, that is to say its IP address.

From DNS resolution, either the DNS server is hosting the requested EID in a domain name format, and so answers to it, or it doesn't have it and sends a response of non-existence.

The major difference with traditional DNS resolution is that by using chord one doesn't want to list all the different crypto-IDs or IDs into the DNS. Furthermore, Chord is based on sorting IDs, because data corresponding of one ID are hosted on the server whose ID is next to it.

The security extension of DNS, DNSSEC, is sorting DNS domain name for proof-of-non-existence purpose. The RRset NSEC works like a dictionary indicating the previous and the next domain name. Thanks to the sorting rule, if a domain name is not between the previous and the next one, you know it does not exist, and the requested domain name is not hosted by this server.

Our architecture requires that DNS servers implement DNSSEC. The DNS server should host at least one chord server. When the request for crypto-ID *2* is received by the DNSSEC server, the DNSSEC cannot provides the answers since it does not host the EID. Thanks to the NSEC RRset, the DNSSEC server is sending the ID of the next closest chord node ID registered in the DNS file, i.e. in our case *3*.

According to the DNSSEC configuration file in figure 5, as a request to *2.company.com*, the resolver will receive an answer indicating that the *2.company.com* is not hosted by the DNSSEC server. The fields indicating this proof of non-existence will be "`1.company.com 8640 IN NSEC 3.company.com RRSIG CHORDSERVER NSEC`". The resolver then knows it has to send a chord request to the chord node

```
0.company.com        8640  IN   CHORDNODE   "192.234.98.88"
"company.com."
                     8640  IN   RRSIG       CHORDNODE [...]
                     8640  IN   NSEC        1.company.com RRSIG
CHORDNODE NSEC
                     8640  IN   RRSIG       NSEC [...]

1.company.com        8640  IN   CHORDNODE   "192.234.98.99"
"company.com."
                     8640  IN   RRSIG       CHORDNODE [...]
                     8640  IN   NSEC        3.company.com RRSIG
CHORDNODE NSEC
                     8640  IN   RRSIG       NSEC [...]
3.company.com        8640  IN   CHORDNODE   "192.234.98.100"
"company.com."
                     8640  IN   RRSIG       CHORDNODE [...]
                     8640  IN   NSEC        0.company.com RRSIG
CHORDNODE NSEC
                     8640  IN   RRSIG       NSEC [...]
```

Fig. 5. Mixed DNS CHORD DNSSEC file zone

3.company.com. This IP address will be either indicated into the `ADDITIONAL DATA` section of the answer, or will need a specific DNS query.

6.2 Discussion on Performance and Security

DNS performance is very high due to the simple and scalable delegation mechanisms and the cache mechanism. The proposed mixed architecture is using DNSSEC which has different performances then DNS, and Figure 6 pointed out measured differences between DNS and DNSSEC on update operations.

On the other hand chord resolution performance is low due to the routing algorithm (Log(N)). [9]indicates there are no major advantages in term of performances to migrate from DNS architecture to a peer-peer DHT one, and performance of DHT can only compete DNS ones with additional mechanisms. Using the DNSSEC protocol enables the client to reduce all chord requests and get directly the proper chord node.

This is a valuable way to enhance and improve performance of the traditional chord resolution, and one can assume that the overall resolution performance is similar as a DNSSEC resolution. Of course the last assumption is that the DNSSEC files are optimized and that most of the chord nodes are hosted in those files. This optimization only impacts the naming resolution performance. In fact, a chord resolution can be handled with any chord node, but routing discovery is then required. This property enables us to believe that the use of chord and DNSSEC can be quite robust to bad configuration. We also believe that this architecture with a delegation from DNS to chord might bring solution to dynamic update. As showed in Figure 6 DNSSEC update performances are very low with DNSSEC. This is mainly due to the fact that adding/deleting a data to a DNSSEC file zone requires the server to re-sign the zone. This architecture requires that DNSSEC files hosts relatively static data that are basically the locators of the chord servers. In that sense dynamic updates aspects are delegated to the chord part of the architecture, and have no impact on the DNSSEC part of the architecture.

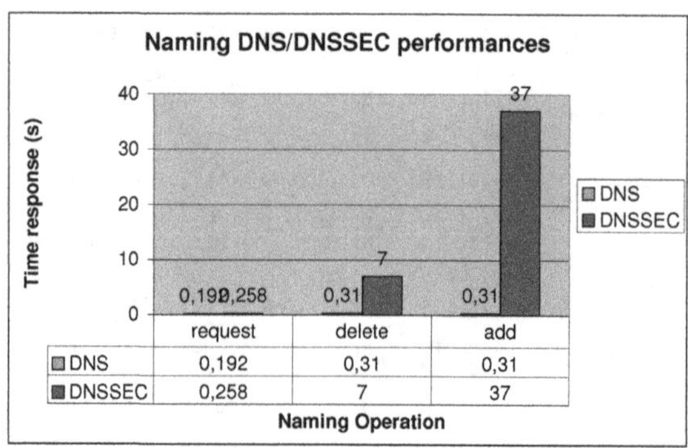

	request	delete	add
DNS	0,192	0,31	0,31
DNSSEC	0,258	7	37

Fig. 6. DNSSEC performances issues. Delete and add operations are proceeded for 500 RRDATA.

In term of security, DNSSEC is securing the DNS part of the architecture. Data are signed, and the chain of trust is built until the domain name that is redirecting to the DHT database.

7 Conclusion

This paper provides a way to manage cryptoID by creating a new EID format. This EID is associating a domain name to the cryptoID. The naming architecture takes advantage of DNS by creating communities as independent administrative domain, considering a single Naming entry point and a chain of trust with the future DNSSEC deployment. On the other hand DHT enables management of flat data like cryptoIDs, and provides scalability, robustness as well as easy domain to administrate. More then the combination of two naming architecture, the proposed mixed architecture is taking advantage of this combination to enhance the DHT resolution with the use of DNSSEC.

This architecture is also an architecture that can be considered as a transition naming architecture before considering EIDs only composed of flat data like cryptoIDs. Those EIDs would then be right under the root domain (.) or would be considered to belong to the root community.

Although the LISP architecture requires a new infrastructure of the core network, or at least is affecting the routing aspects, the architecture proposed in this paper is only considering the naming function with a DNS/DHT resolver on the client side. Nevertheless this is done at the expense of modifying the EID format. Impacts of modifying the EID format should clearly be understood especially when migration is concerned. At least we should consider if there are possibilities to specify the EID type in the header. This would for instance require no additional core network architecture modification, and would enable nodes to use self generated identities which matches the ad-hoc purposes.

If EID needs to match the IP address format, then a hash function can be used to generate EID with crypto-ID and FQDN as input. The NERD [15] or CONS [14] architecture might then be still required for EID-to-RLOC resolution or EID-to-(FQDN, crypto-ID) resolution.

Deployment issues should also investigate on how to deploy the DHT resolver. Migration could then consider alternative issue by mapping DHT with DNS interfaces like HOLD [17]. This would enable DNS-only resolver to perform EID resolution.

Analyses to see whether new format of EIDs brings complexity or rather offers new possibilities / flexibility have to be looked at carefully. Then implementation and testing this architecture should be provided and compared to the one proposed in LISP.

Acknowledgment

We would like to thank all people that are involved in the Naming Theme in the Ambient network project, but our special thinking go especially to Christopher Reichert highly committed in the project, who passed away on October 2006.

References

1. Huston, G.: Architectural Commentary on Site Multi-homing using a Level 3 Shim (January 2005)
2. Moskowitz, R., Nikander, P.: Host Identity Protocol (HIP) Architecture. RFC 4423 (Informational) (May 2006)
3. Kaashoek, F., Karger, D., Morris, R., Sit, E., Stribling, J., Brunskill, E., Cox, R., Dabek, F., Li, J., Muthitacharoen, A., Stoica, I.: Chord
4. Arends, R., Austein, R., Larson, M., Massey, D., Rose, S.: DNS Security Introduction and Requirements. RFC 4033 (Proposed Standard) (March 2005)
5. Arends, R., Austein, R., Larson, M., Massey, D., Rose, S.: Resource Records for the DNS Security Extensions. RFC 4034 (Proposed Standard) (March 2005)
6. Arends, R., Austein, R., Larson, M., Massey, D., Rose, S.: Protocol Modifications for the DNS Security Extensions. RFC 4035 (Proposed Standard) (March 2005)
7. Ramasubramanian, V., Sirer, E.G.: Beehive: O(1)lookup performance for power-law query distributions in peer-to-peer overlays. In: NSDI 2004. Proceedings of the 1st conference on Symposium on Networked Systems Design and Implementation, Berkeley, CA, USA, USENIX Association, p. 8 (2004)
8. Ramasubramanian, V., Sirer, E.G.: The design and implementation of a next generation name service for the internet. In: SIGCOMM 2004. Proceedings of the 2004 conference on Applications, technologies, architectures, and protocols for computer communications, pp. 331–342. ACM Press, New York (2004)
9. Pappas, V., Massey, D., Terzis, A., Zhang, L.: A Comparative Study of the DNS Design with DHT-Based Alternatives (March 2005)
10. Stoica, I., Morris, R., Karger, D., Kaashoek, F.M., Balakrishnan, H.: Chord: A scalable peer-to-peer lookup service for internet applications. In: SIGCOMM 2001. Proceedings of the 2001 conference on Applications, technologies, architectures, and protocols for computer communications, vol. 31, pp. 149–160. ACM Press, New York (2001)

11. Cox, R., Muthitacharoen, A., Morris, R.T.: Serving DNS using a Peer-to-Peer Lookup Service (March 2002)
12. Farinacci, D., Fuller, V., Oran, D., Meyer, D.: Locator/ID Separation Protocol (LISP)(July 2007)
13. Jen, D., Meisel, M., Massey, D., Wang, L., Zhang, B., Zhang, L.: APT: A Practical Transit Mapping Service (July 2007)
14. Brim, S., Chiappa, N., Farinacci, D., Fuller, V., Lewis, D., Meyer, D.: LISP-CONS: A Content distribution Overlay Network Service for LISP (July 2007)
15. Lear, E.: NERD: A Not-so-novel EID to RLOC Database (July 2007)
16. Doi, Y.: Dns meets dht: Treating massive id resolution using dns over dht. In: Saint, pp. 9–15 (2005)
17. Considine, J., Walfish, M., Andersen, D.G.: A pragmatic approach to dht adoption. Technical report

An Efficient and Reliable Anonymous Solution for a Self-organized P2P Network

J.P. Muñoz-Gea, J. Malgosa-Sanahuja, P. Manzanares-Lopez, J.C. Sanchez-Aarnoutse, and J. Garcia-Haro

Department of Information Technologies and Communications
Polytechnic University of Cartagena
Campus Muralla del Mar, 30202, Cartagena, Spain
{juanp.gea,josem.malgosa,pilar.manzanares,juanc.sanchez,
joang.haro}@upct.es

Abstract. In this paper, a new mechanism to achieve anonymity in peer-to-peer (P2P) file sharing systems is proposed. As usual, anonymity is obtained by means of connecting the source and destination peers through a set of intermediate nodes, creating a multiple-hop path. The main contribution of the paper is a distributed algorithm able to guarantee the anonymity even when a node in a path fails (voluntarily or not). The algorithm takes into account the inherent costs associated with multiple-hop communications and tries to reach a well-balanced solution between the anonymity degree and its associated costs. Some parameters are obtained analytically but the main network performances are evaluated by simulation. We quantify the costs associated with the control packets used by the distributed recovery algorithm. On the other hand, we also measure the anonymity provided by our system (benefit), using a simulation-based analysis to calculate the average entropy.

1 Introduction

P2P networks are the most popular architectures for file sharing. In some of these scenarios, users are also interested in keeping mutual anonymity; that is, any node in the network should not be able to know who is the exact origin or destination of a message. Traditionally, due to the connectionless nature of IP datagrams, the anonymity is obtained by means of connecting the source and destination peers through a set of intermediate nodes, creating a multiple-hop path between the pairs of peers.

There are various existing anonymous mechanisms with this operation, but the most important are the mechanisms based in Onion Routing [1] or Crowds [2]. The differences between them is the way paths are determined and packets are encrypted. In Onion Routing, packets are encrypted according to a predetermined path before they are submitted to the network; in Crowds, a path is configured as a request traverses the network and each node encrypts the request for the next member on the path.

The above procedure to reach anonymity has two main drawbacks. The first one is that the peer-to-peer nature of the network is partially eliminated since

D. Medhi et al. (Eds.): IPOM 2007, LNCS 4786, pp. 25–36, 2007.

now, the peers are not directly connected, since there is a path between them. Therefore, the costs of using multiple-hop paths to provide anonymity are extra bandwidth consumption and an additional terminal (node) overload (routing connections not created by itself). In addition, as it is known, the peer-to-peer elements are prone to unpredictable disconnections. Although this fact always affects negatively the system performance, in an anonymous network it is a disaster since the connection between a pair of peers probably would fail although both peers are running. Therefore, a mechanism to restore a path when an unpredictable disconnection arises is needed, but it also adds an extra network overhead in terms of control traffic.

Wright *et al.* [3] presented a comparative analysis about the anonymity and overhead of several anonymous communication systems. This work presents several results that show the inability of protocols to maintain high degrees of anonymity with low overhead in the face of persistent attackers. In [4], authors calculate the number of appearances that a host makes on all paths in a network that uses multiple-hop paths to provide anonymity. This study demonstrates that participant overhead is determined by number of paths, probability distribution of the length of path, and number of hosts in the anonymous communication system.

In this paper we propose an anonymity mechanism, for a hybrid P2P network presented in a previous work [5], based on Crowds to create the multiple-hops path. In order to restrict the participant overload, our mechanism introduces a maximum length limit in the path creation process. The main paper contribution is a distributed algorithm to restore a path when a node fails (voluntarily or not). The algorithm takes into account the three costs outlined above in order to obtain an equilibrated solution between the anonymity degree and its associated costs. The main parameters are evaluated analytically and by simulation.

The remainder of the paper is as follows: Section 2 summarizes the main characteristics of our hybrid P2P architecture. Sections 3 and 4 deeply describe the mechanism to provide anonymity. Section 5 shows the simulation results and finally, Section 6 concludes the paper.

2 A Hybrid P2P Architecture

One of the main motivations to use a SuperPeer model in P2P file-sharing applications is the search efficiency. With the solution based in SuperPeers, users submit broad queries to a SuperPeer in order to find data of users' interest. If this SuperPeer cannot find a solution, the query will be forwarded to a limited number of SuperPeers. The overlay network which connects SuperPeers among them can have several topologies: a random network, a DHT, or other topology.

In our proposal [5], we implement the communication among SuperPeers using a hierarchical topology of nodes. With this mechanism, the level-1 SuperPeers add and publish the data owned by a group of usual hosts. The rest of SuperPeers levels do not need to maintain this information. In general, a SuperPeer belonging

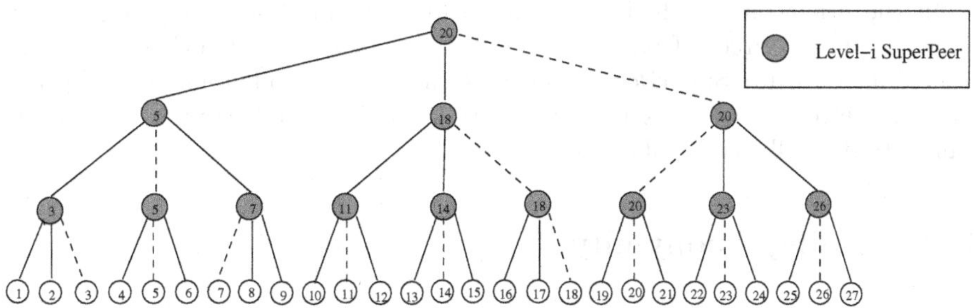

Fig. 1. Proposed system architecture, with a three levels SuperPeer hierarchy. Dashed lines denote the nodes that are SuperPeers in the next level.

to level n is in charge of administering a group of level-$n-1$ SuperPeers, and it can forward the data requests to them if need be. Figure 1 describes the general architecture of the system.

For example, node 1, in order to search for a content, will send the search parameters to its level-1 SuperPeer (node 3), and this one will return information about who has the content in its subgroup. If the search fails, the node 3 will send the request to its associated higher level SuperPeer (node 5). Then, this node will also forward the request to node 7. If it is necessary, this process will continue up to the highest level SuperPeer (node 20).

In addition, our architecture presents two key services: the distributed topology creation service, and the fault-tolerance service. This services use the DHT (distributed hash table) service provided by a structured P2P network. This interface implements a simple store and retrieve functionality, where a *value* is always stored at the live node to which a *key* is mapped by the structured network. Concretely, our network uses as key the $SubgroupID$, that represents a subgroup of SuperPeers or usual hosts, managed by the same higher level SuperPeer; and as value the *IP address* of this higher SuperPeer.

Therefore, when a node wants to join the network, it must firstly contact with an existing usual host and obtain its corresponding $SubgroupID$. Next, using the DHT service it will obtain the IP address of the corresponding SuperPeer. Them, the new node must contact that SuperPeer, which will accept its union if the subgroup is not full. However, if there is no room, the new node will be asked to create a new (randomly generated) subgroup, or it will be asked to join the subgroup that the requested SuperPeer urged to create previously. If a new subgroup is created, the corresponding level-1 SuperPeer will try to join the level-2 subgroup of the previously contacted SuperPeer, or the existing SuperPeer will create a new level-2 subgroup, if there is not any. The level-2 subgroups also have a maximum number of nodes, and therefore the subgroup creation process continues in a similar way up to the highest level of hierarchy. With this algorithm, only the set of level-1 SuperPeer is candidate for being SuperPeer in the rest of levels.

On the other hand, when a new node finds its final subgroup SuperPeer, it notifies its resources of bandwidth and CPU. Thus, the SuperPeer forms an ordered list of future SuperPeer candidates: the longer a node remains connected (and the better resources it has), the better candidate it becomes. This list is transmitted to all the members of the subgroup.

3 Providing Anonymity

In this section we propose a file sharing P2P application built over the previous network architecture which provides mutual anonymity. As it is defined in [6], a P2P system provides mutual anonymity when any node in the network, neither the requester node nor any participant node, should not be able to know with complete certainty who is the exact origin and destination of a message.

Our solution defines three different stages. On one hand, all peers publish their contents within their local subgroups (publishing). To obtain the anonymity during this phase, a random walk procedure will be used. On the other hand, when a peer initiates a complex query, a searching phase will be carried out firstly (searching). Finally, once the requester peer knows all the matching results, the downloading phase will allow the download of the selected contents (downloading). In any case, the anonymity is always maintained even under a failure pattern (a common situation in P2P systems). In the following section these three stages are described in detail.

3.1 Publishing the Contents

To maintain the anonymity, our solution defines a routing table at each peer and makes use of a random walk (RW) technique to establish a random path from the content source to the subgroup SuperPeer. When a node wants to publish its contents, first of all it must choose randomly a connection identifier (great enough to avoid collisions). This value will be used each time this peer re-publishes a content or publishes a new one.

The content owner randomly chooses another active peer (a peer knows all the subgroup members, who are listed in the future SuperPeer candidate list, see Section 2) and executes the RW algorithm to determine the next peer in the RW path as follows: The peer will send the *publish message* directly to the subgroup SuperPeer with probability $1-p$, or to the randomly chosen peer with probability p. This message contains the connection identifier associated to the content owner and the published content metadata. The content owner table entry associates the connection identifier and the next peer in the RW path. Each peer receiving a *publish message* follows the same procedure. It stores an entry in its table, which associates the connection identifier, the last peer in the RW path and the next node in the RW path (which will be determined by the peer using the RW algorithm as it was described before).

To prevent a message forwarding loop within the subgroup, each initial *publish message* (that is generated by the content owner) is attached with a TTL (Time

To Live) value. A peer receiving a *publish message* decrements the TTL by one. Then, if the resulting TTL value is greater than 1, the peer executes the RW algorithm. Otherwise, the message is sent directly to the subgroup SuperPeer. When the *publish message* is received by the subgroup SuperPeer, it just stores the adequate table entry that associates the connection identifier, the published content metadata and the last peer in the RW path.

Thanks to the publishing procedure, each subgroup SuperPeer knows the metadata of the contents that have been published into its subgroup. However, this publishing procedure offers sender anonymity: Any peer receiving a *publish message* knows who the previous node is, but it does not know if this previous peer is the content owner or just a simple message forwarder.

With the above pocedure, the probability that the RW has n hops is

$$P(n) = \begin{cases} p^{n-1}(1-p) & n < TTL, \\ p^{TTL-1} & n = TTL. \end{cases} \tag{1}$$

Therefore the mean length of a RW is

$$\overline{RW} = \frac{1-p^{TTL}}{1-p} \xrightarrow[p \to 1]{} TTL. \tag{2}$$

We can also derive an expression for the number of forwarding paths on a participant node. If we suppose that in a subgroup there are N participant nodes and P paths, we can use the analysis presented in [4] to estimate the expectation for the number of forwarding paths (F_j) on an arbitrary participant v_j $(0 \leq j < N)$

$$\overline{F_j} = \left(\frac{P}{N}\right) \overline{RW} = \overline{RW} \xrightarrow[p \to 1]{} TTL. \tag{3}$$

since in our architecture N is equal to P.

Equations 2 and 3 show that, in our proposal, the associated costs with the anonymity (extra bandwidth consumption and nodes overload) are limited, independently of p, by the TTL value. As it will see, an appropriate value for the TTL allows a good tradeoff between the costs-benefits associated with the anonymity.

The RW path distributely generated in the publishing phase will be used during the searching and downloading phases, as it will be described later. Therefore, it is necessary to keep updated the RW paths according to the non-voluntary peer departures. From its incorporation to the anonymous system, each peer maintains a *timeout timer.* When its timeout timer expires, each peer will check the RW path information stored in its table. For each entry, it will send a *ping message* to the previous and the following peer. If a peer detects a connection failure, it generates a *RW failure notification* that is issued to the active output and peer by peer to the subgroup SuperPeer (or to the content owner) using the RW path. Each peer receiving a *RW failure notification* deletes the associate table entries (only one entry in intermediate peers, and one entry by each published content in the subgroup SuperPeer peer). In addition, the content owner

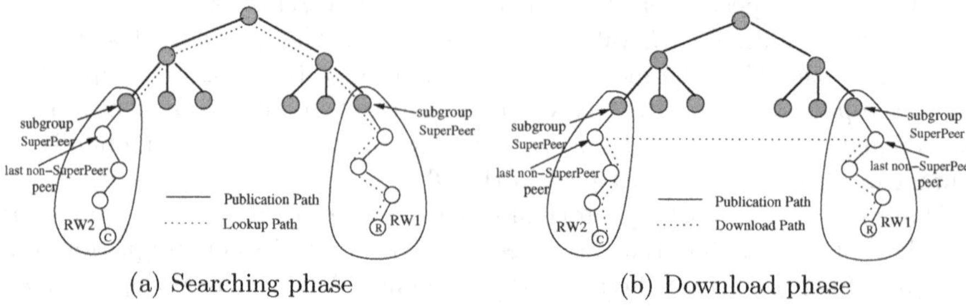

Fig. 2. Providing anonymity

re-publishes all its contents. Notice that, any other peer in the RW not aware about the failure will detect it when its own timer expires. On the other hand, if an intermediate peer wants to leave the system in a voluntary way, it will get in touch with the previous and the following peer associated to each table entry in order to update their entries.

An additional verification is carried out each time a new content is going to be published. Before sending the *publish messages*, the content owner peer makes a *totalPing* (a ping sent to the subgroup SuperPeer through the RW path). If the *totalPing* fails (any intermediate peer is disconnected), the content owner peer re-publishes all its contents initiating the creation of a new distributed RW path. The intermediate peers belonging to the failed RW path will update their table entries when their timeout expire.

3.2 Searching the Contents

The searching procedure to obtain requester anonymity is described in this section. Figure 2(a) describes this procedure. When a requester (R) wants to make a complex search, first of all it must select its connection identifier. If the peer already chose a connection identifier during a publish phase, this value is used. Otherwise, a connection identifier is chosen[1].

Then, the peer generates a new type of message called *lookup message*. A *lookup message*, that encapsulates the metadata of the searched content and the connection identifier associated to the requester peer (R), is forwarded towards the local subgroup SuperPeer using the RW path associated to this connection identifier (RW1). If the connection identifier is new, the RW1 path from the requester peer towards the subgroup SuperPeer is generated, distributely stored, and maintained as it was explained in the previous section.

Once the *lookup message* has been received by the local subgroup SuperPeer, it is able to locate the matching results within its subgroup. On the other hand, if this local matching fails, our system uses the SuperPeers hierarchy to distribute the lookup messages until finding the searched content.

[1] Each peer is associated to only one connection identifier.

Each SuperPeer with a positive matching result must answer the requesters subgroup SuperPeer with a *lookup-response message*. This message contains the connection identifier established by the requester peer, the matching content metadata, the connection identifier established by the content owner peer and the identity of the next peer in the RW path (RW2) towards the content owner (C).

In fact, before sending the *lookup-response message*, the subgroup SuperPeer will check the complete RW2 path towards the content owner by means of a *totalPing*. If the RW2 path fails, the subgroup SuperPeer will send to all the peers in its subgroup a broadcast message containing the associate connection identifier. The peer that established this connection will republish all its contents (that is, it initiates the creation of a new RW for the connection identifier). Peers that maintain a table entry associated to this connection identifier will delete it. The rest of peers will take no notice of the message. Consequently, the subgroup SuperPeer only sends a *lookup-response message* to the requester subgroup SuperPeer after checking that the associate RW2 path is active.

Finally, the *lookup-response message* will be forwarded through the requester subgroup towards the requester peer using the adequate and distributed RW1 path stored in the tables.

3.3 Downloading the Contents

Once the requester peer receives the *lookup-response* messages, it is able to select which contents it wants to download. The anonymity is also important during this third phase: both the requester peer and the content owner peer should keep anonymous to the rest of peers in the system.

In the download phase (see Figure 2(b)), the requester peer (R) generates a *download-request message* which initially encapsulates the required content metadata, the connection identifier established by itself, the connection identifier established by the content owner (C) and the identity of the last non-SuperPeer peer in the RW2 (in the content subgroup). This message will be forwarded towards the subgroup SuperPeer using the distributely stored RW1 path. However, this message will not reach the subgroup SuperPeer. The last non-SuperPeer peer in the RW1 path (the previous to the SuperPeer) will be the responsible for sending the *download-request message* to the last non-SuperPeer peer in the RW2 (in the content subgroup). Before sending the message to the content owner subgroup, this peer (the last non-SuperPeer peer in RW1) must encapsulate an additional value: its own identity. The message will be forwarded through the content owner subgroup using the distributely stored RW2 path until it reaches the content owner.

Finally, the content owner peer will start the content delivery. The *download messages* encapsulate the requited content, the connection identifier established by the requester peer, the connection identifier established by itself and the identity of the last non-SuperPeer peer in the RW1 (in the requester subgroup). These messages will be forwarded towards the subgroup SuperPeer using the adequate RW2 path. However, they will not reach the subgroup SuperPeer. The last non-SuperPeer peer in the RW2 path will be the responsible for sending them to the node indicated in the message. Once in the requester subgroup, the messages will be forwarded using the RW1 path until they reach the requester peer. Therefore, the mean length of a download path (DP) is

$$\overline{DP} = 2(\overline{RW} - 1) + 1 = 2\frac{1 - p^{TTL}}{1 - p} - 1 \tag{4}$$

Note that any SuperPeer do not participate during the download process, making lighter their processing load.

4 Dynamic Aspects of the System

The peer dynamism or *churn* is an inherent feature of the P2P networks. Our solution exploits this property to improve the anonymity behaviour of the system. As it has been described in the previous section, each peer establishes a RW path towards its subgroup SuperPeer during the publishing or the searching phases. If there was no churn, these RW paths (which are randomly established) would not change as the time passes.

To optimize the system behaviour when a subgroup SuperPeer changes we propose some actions. If a subgroup SuperPeer changes voluntarily (another peer in the subgroup has become a most suitable SuperPeer or the SuperPeer decides to leave the system), it will send to the new SuperPeer a copy of its table. On the other hand, to solve the non-voluntary SuperPeer leaving, we propose to replicate the table of the subgroup SuperPeer to the next two peers in the future SuperPeer candidate list. Consequently, when the subgroup SuperPeer fails, the new subgroup SuperPeer will be able to update its table taking into account the published contents.

5 Simulations

We have proposed a mechanism to provide anonymity to a hybrid P2P architecture. This mechanism works fine under a join/leave or failure scenario, but it involves an extra control packet interchange. It is necessary to quantify this cost. On the other hand, it is also necessary to calculate the anonymity provided by our system.

We have developed a discrete event simulator in C language to evaluate our system. At the beginning, the available contents are distributed in a random way among all the nodes of the network. The node life and death times follow a Pareto distribution using $\alpha = 0.83$ and $\beta = 1560$ sec. These are usual values for these parameters (as it is exposed in [7]). The time between queries follows an exponential distribution with mean 3600 sec. (1 hour), and if a query fails the user makes the same request for three times, with an empty interval of 300 sec. among them. The *timeout timer* has a value of 300 sec. In the path construction algorithm we use the following values: $p = 0.75$ and $TTL = 10$.

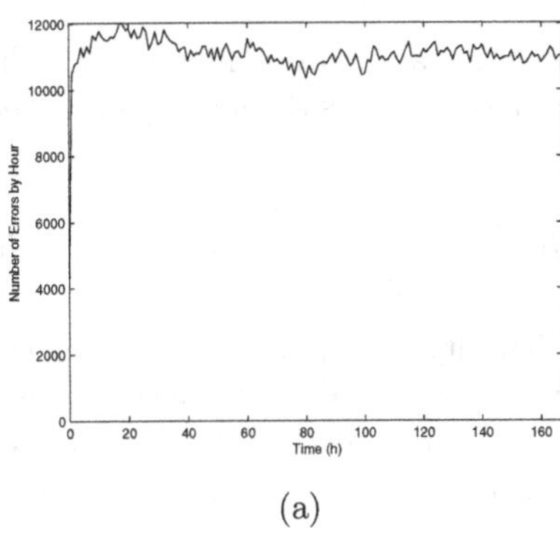

(a)

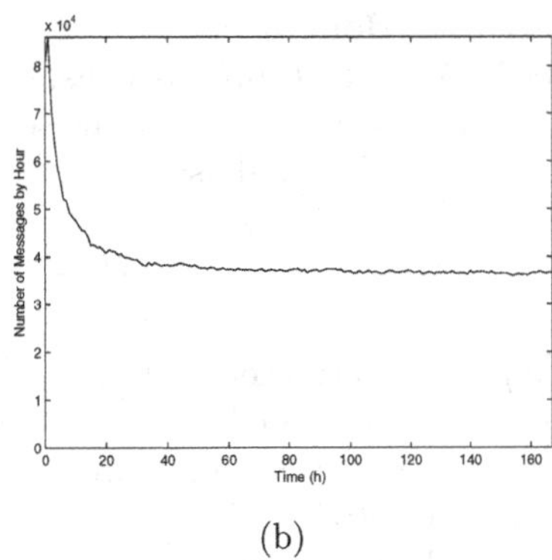

(b)

Fig. 3. (a) Number of request that cannot be carry out in an hour because at least one node in the download path is out. (b) Average number of control messages in function of time.

Finally, the simulation scenario is formed by 12,500 different contents uniformly distributed among 6,250 nodes. The sub-group sizes are set to 50 and 5 nodes for the first and the rest of the hierarchy respectively. Therefore, there are 125 level-1 SuperPeers, 25 level-2, 5 level-3 and one level-4.

In the results represented in Figure 3(a), our system doesn't implement the reliability mechanism. It shows the number of requests that cannot be carry out in an hour because, at least one node in the download path is down when the download is performed. This result is represented in function of time (in hours) and it varies around 11,000 failed requests. Therefore, it is clear than in a real anonymous P2P network (with unpredictable node failures) it is necessary to supply a mechanism to reconstruct efficiently the paths, although it entails an extra network overhead in terms of control packets. This is called the anonymity cost.

Anonymity Costs

In our network, a download process never fails since it is a reliable anonymous P2P network, but this feature involves extra control traffic. Figure 3(b) represents the average number of control messages in function of time. Initially, due to simulation constraints, the *timeout timer* expires simultaneously in a lot of nodes, but in steady state the average number of control messages is under 40,000. Usually, control messages have a small payload (about 10 bytes) and a IP/TCP header (40 bytes). Therefore, if we suppose each control message has a length of 50 bytes, control traffic supposes only a 4.4 kbps traffic rate.

Furthermore, as we said in Section 3, the other associated costs with the anonymity (extra bandwidth consumption and nodes overload) are limited independently of p, by the TTL value. Therefore, an appropiate value for the TTL allows a good trade-off between the costs-benefits associated with the anonymity.

Anonymity Benefits

In 2002, Díaz *et al.* proposed using entropy as an anonymity metric [8]. The degree of anonymity of a user who sends a message that goes through a corrupted path (H_c) is calculated as:

$$H_c = -\sum_i p_i \cdot log_2 p_i \tag{5}$$

where p_i is the attacker's estimate of the probability that participant i was the sender of a intercepted message. However, we have to take into account that the message could go only through honest nodes, and in this case the degree of anonymity (H_h) is the maximum, since in this case p_i is equally distributed among the honest nodes:

$$H_h = -\sum_{i=1}^{n-c} \frac{1}{n-c} log_2 \left(\frac{1}{n-c} \right) = log_2(n-c) \tag{6}$$

being n the total number of nodes and c the total number of compromised nodes (attacker who can collaborate among them). If the first case occurs with probability p_c and the second with probability p_h, the average degree of anonymity of the system (H) is:

$$H = p_c H_c + p_h H_h \tag{7}$$

We can interpret this in another way. Let us introduce two random variables: X, modelling the senders ($X = 1, ..., n-c$), and Y, modelling the predecessor observed by the attacker nodes ($Y = 1, ..., n-c, \oslash$; where $\oslash$ represents that there is not an attacker in the path). With this definition we can redefine expression 7 as,

$$\begin{aligned} H &= p_c H_c + p_h H_h = \\ &= \sum_{y \geq 1} Pr[Y = y] H(X|Y = y) + Pr[Y = \oslash] H(X|Y = \oslash) = \\ &= \sum_y Pr[Y = y] H(X|Y = y) \end{aligned} \tag{8}$$

We use the simulation-based analysis proposed in [9] to calculate H, as follows. We keep a counter $C_{x,y}$ for each pair of x and y. We pick a random value x for X and simulate a request. We forward the request according to the algorithm until it either is sent to the destination or is intercepted by a corrupt node. In the former case, we write down the observation $y = \oslash$, while in the latter case we set y to the number of the predecessor node. In both cases, we increment $C_{x,y}$. Next, we can compute

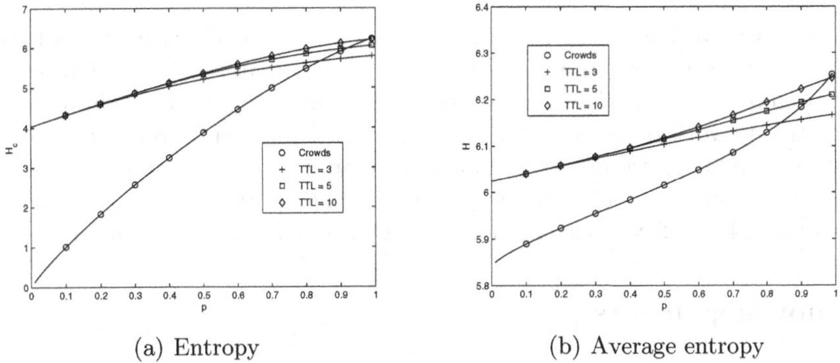

(a) Entropy (b) Average entropy

Fig. 4. Degree of anonymity

$$H(X|Y=y) = -\sum_{x} q_{x,y} log_2 q_{x,y} \tag{9}$$

where $q_{x,y} = C_{x,y}/K_y$, and $K_y = \sum_x C_{x,y}$.

We have to take into account that the distribution $q_{x,y}$ is only an estimate of the true distribution p_i. Therefore, in first place, we have to get the appropriate number of samples in order to get a good approximation for H (a sample or experiment is the selection by a non-corrupted node, i.e. belonging to X, of a random walk).

We are going to calculate H in a typical scenario: every subgroup has 100 nodes, and 10 of them are corrupted nodes. In the path construction algorithm we use the following values: $p = 0.75$ and $TTL = 5$. We observe that from 10,000 samples the entropy stabilizes up to the second decimal.

Next, it is showed the influence of p and TTL in the degree of anonymity (entropy (H_c) and average entropy (H)) of our system, and it is compared with the degree of anonymity provided by Crowds (presented in [8]). We have to remember that in Crowds there is not a limit for the path length, and there is only a path among the initiator and the responder. In our network, paths have a length limited by TTL, and the download path (among the initiator an the responder) is composed by two publication paths.

If we concentrate in Figure 4(a) we observe that if $p < 0.8$ the entropy provided by our network is bigger enough to the entropy provided by Crowds, independently of the value of TTL. If the TTL value increases to 5 the entropy in our network is greater if $p < 0.9$, and with $TTL = 10$ the entropy in our network is always greater or equal to the entropy provided by Crowds.

For the average entropy the results are equivalent. With $TTL = 10$ the average entropy in our network is always greater or equal to the average entropy provided by Crowds. But, with a $p < 0.8$ and small TTL, the average entropy in our network is bigger to Crowds.

6 Conclusions

In this paper, we have proposed a reliable anonymity mechanisms for a hybrid P2P network presented in a previous work. Unfortunately, the mechanisms to provide anonymity entails extra bandwidth consumption, extra nodes overload and extra network overhead in terms of control traffic. However, our proposal achieves a great degree of mutual anonymity only with a minimum bandwidth consumption, corresponding to both a limited control traffic rate and a limited path length. In addition, the anonymity is guarateed under node failures.

Acknowledgements

This work has been funded by the Spanish Research Council under the ARPaq (TEC2004-05622-C04-02/TCM) and the CON-PARTE-1 (TEC2007-67966-C03-01/TCM) projects. Juan Pedro Muñoz-Gea also thanks the Spanish MEC for a FPU (AP2006-01567) pre-doctoral fellowship.

References

1. Reed, M.G., Syverson, P.F., Goldshlag, D.M.: Anonymous connections and onion routing. IEEE Journal on Selected Areas in Communications 16(4), 482–494 (1998)
2. Reiter, M.K., Rubin, A.D.: Crowds: Anonymity for web transactions. Communications of the ACM 42(2), 32–48 (1999)
3. Wright, M., Adler, M., Levine, B.N., Shields, C.: An analysis of the degradation of anonymous protocols. In: NDSS 2002. Proceedings of the Network and Distributed Security Symposium, San Diego, CA, USA (2002)
4. Sui, H., Chen, J., Chen, S., Wang, J.: Payload analysis of anonymous communication system with host-based rerouting mechanism. In: ISCC 2003. Proceedings of the Eighth IEEE International Symposium on Computers and Communications, Kemer-Antalya, Tuerkey, IEEE Computer Society Press, Los Alamitos (2003)
5. Muñoz-Gea, J.P., Malgosa-Sanahuja, J., Manzanares-Lopez, P., Sanchez-Aarnoutse, J.C., Garcia-Haro, J.: A self-organized p2p network for an efficient and secure content location & download. In: Moro, G., Bergamaschi, S., Joseph, S., Morin, J.-H., Ouksel, A.M. (eds.) DBISP2P 2006/2005. LNCS, vol. 4125, pp. 368–375. Springer, Heidelberg (2007)
6. Pfitzmann, A., Kohntopp, M., Showtack, A.: Anonymity, unlinkability, unobservability, pseudonymity and identity management - A consolidated proposal for terminology. Manuscript (May 2006)
7. Li, J., Stribling, J., Morris, R., and Kaashoek, M. F.: Bandwidth-efficient management of dht routing tables. In: NSDI 2005. Proceedings of the 2nd USENIX Symposium on Networked Systems Design and Implementation, Boston, MA, USA (2005)
8. Díaz, C., Seys, S., Claessens, J., Preneel, B.: Towards measuring anonymity. In: Dingledine, R., Syverson, P.F. (eds.) PET 2002. LNCS, vol. 2482, Springer, Heidelberg (2003)
9. Borisov, N: Anonymous routing in structured peer-to-peer overlays, Ph.D. Thesis, UC Berkeley, May (2005)

Analysis of Diagnostic Capability for Hijacked Route Problem

Osamu Akashi[1], Kensuke Fukuda[2], Toshio Hirotsu[3], and Toshiharu Sugawara[4]

[1] NTT Network Innovation Labs., Midori-cho 3-9-11, Musashino-shi, Tokyo, Japan
akashi@core.ntt.co.jp
[2] National Institute of Informatics, Chiyoda Tokyo, Japan
kensuke@nii.ac.jp
[3] Toyohashi University of Technology, Aichi, Japan
hirotsu@ics.tut.ac.jp
[4] Waseda University, Shinjuku Tokyo, Japan
sugawara@waseda.jp

Abstract. Diagnosis of anomalous routing states is essential for stable inter-AS (autonomous system) routing management, but it is difficult to perform such actions because inter-AS routing information changes spatially and temporally in different administrative domains. In particular, the route hijack problem, which is one of the major routing-management issues, remains difficult to analyze because of its diverse distribution dynamism. Although a multi-agent-based diagnostic system that can diagnose a set of routing anomalies by integrating the observed routing statuses among distributed agents has been successfully applied to real Internet service providers, the diagnostic accuracy depends on where those agents are located on the BGP topology map. This paper focuses on the AS adjacency topology of an actual network structure and analyzes hijacked-route behavior from the viewpoint of the connectivity of each AS. Simulation results using an actual Internet topology show the effectiveness of an agent-deployment strategy based on connectivity information.

1 Introduction

The Internet consists of more than 20,000 autonomous systems (ASs) that correspond to independent network management units such as Internet service providers (ISPs). The inter-AS routing information is locally exchanged among neighboring ASs and distributed through the Internet in a hop-by-hop manner by the border gateway protocol (BGP) [1]. There is no centralized control system and no global view.

However, the actual routing information changes spatially and is temporally affected by various events including human operation errors. The spatial changes easily lead to inconsistent routing states among several ASs, even though each AS is working consistently with respect to its neighboring ASs. Moreover, the ASs experiencing anomalies might not be those causing the anomalies. The temporal

D. Medhi et al. (Eds.): IPOM 2007, LNCS 4786, pp. 37–48, 2007.

changes require verification at multiple observation points on an on-demand basis. Detection and diagnosis require human operators to repeatedly observe and analyze large amounts of routing information, including raw data such as BGP update messages. Operators can use tools such as `ping`, `traceroute`, and `looking glass` [2], but they must use these tools repeatedly over a long period to confirm their own AS's advertisement and find an anomaly as quickly as possible. These can easily lead to unstable inter-AS routing states. This type of BGP information dynamics sometimes makes it meaningless to perform analysis in advance.

Diagnosis based on on-demand observation is inevitable to solve these problems. Moreover, a coordination framework among ASs for monitoring, analyzing, and controlling inter-AS routing is required as essential functions of future Internet design (FIND) for realizing stability and flexibility at the inter-AS level. For example, the route hijack problem, which is one of the major inter-AS routing-management issues, strongly requires cooperative framework among concerning ASs to detect, identify, and mitigate this anomaly.

In practice, a limited number of agents are distributed on the Internet, and more suitable agents for diagnosis should be dynamically sought. For example, if a hijacked prefix is distributed in some area, this anomalous routing state should be found and contaminated area should be identified. An important concern regarding multi-agent system (MAS) approaches is how to ensure the collaboration that can be achieved through many agents appropriately deployed and selected based on their network locations, abilities, and workloads. Recent Internet research indicates that cooperation among simple neighborhood-based agents is neither scalable nor effective. The degree distribution of the real Internet topology is characterized by a power-law [3,4]: It has a small number of hub ASs, each having many direct connections with a huge number of other ASs. Therefore, the agents in hub ASs can play an important role in the distribution of best-path entries. This means that these agents have a higher probability of detecting hijacked routes than agents in ASs having lower connectivity.

Although a lot of work on analyzing network structures and their routing behaviors has been reported, the analysis of the behavior of hijacked routes in relation to filtering ASs from the viewpoint of their connectivity has not been reported as far as we know. Therefore, in this paper, we first focus on the AS adjacency topology of an actual network structure and analyze route distribution and hijacked-route behavior from the viewpoint of the connectivity of each AS. We then discuss these results in relation to multi-agent-based management architectures and then investigate the effectiveness of an agent-deployment strategy through simulation results.

2 Background

As pointed out in [5,6], cooperative distributed problem solving (CDPS) provided by a MAS can adequately deal with such problems because it offers three useful features: 1) CDPS coincides with the control architecture, and monitoring methods should be managed on a request-and-accept basis rather than by using centralized control approaches, 2) observed results that include statistical analysis should be shared after local calculation to improve efficiency and scalability, and 3) operation availability such as message relaying among agents, whose effectiveness was verified through deployment, should be established. In particular, in the case of route hijacking, this message relaying enables an effective communication method even if the observed IP-prefix where the requesting AS is located is hijacked. Approaches for detecting hijacking such as MyASN [7], PHAS [8], and PGBGP [9] have been proposed, but they cannot report this event to systems located in the hijacked prefixes because these systems have no cooperative communication mechanism. The integration of observed results from several ASs can provide more accurate network views for effectively inferring the behaviors of BGP information flows.

2.1 MAS-Based Inter-AS Routing-Anomaly Diagnostic System

As an instance of a cooperative diagnostic system that can cope with these inter-AS routing problems, a MAS-based diagnostic system called ENCORE [10,5] has been developed. This subsection briefly introduces basic cooperative actions of the ENCORE system. Although the analysis in this paper does not necessarily depend on the ENCORE system, we believe these cooperative actions of the really deployed system on the Internet are considered as an essential index.

According to the analysis in [5], a global view of the current routing information that has spread among different administrative domains is essential for diagnosing inter-AS routing anomalies. Since complete understanding of the global view is impossible, we utilize the routing information observed almost simultaneously at multiple ASs. By integrating these observed results, we can infer part of the global view for the purpose of diagnosis. The basic idea of this system is the reflector model, as illustrated in Fig. 1. The essence of this model is to provide a function that enables an agent to request a remote agent to observe routing information about a specific AS, which is usually the AS of the requesting agent. The routing information observed and analyzed by remote agents is sent to the requesting agent.

A relay function is necessary to cooperatively deliver important messages to destination agents even when direct IP routes for message delivery become unavailable, as in the case of a filter setting error or hijacked route anomaly. This function is achieved by having the agents act as application gateways. This function is useful

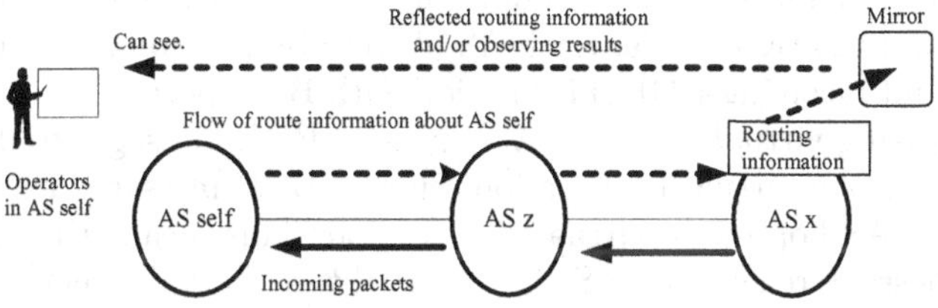

Fig. 1. Reflector model: basic idea for observing spread of information

because 1) the system can use paths that are not used in usual IP routing, and these paths can include non-transit ASs, and 2) messages whose source IP addresses have changed can pass through misconfigured filters with a high probability.

Each agent needs a strategy that defines how to cooperate with other agents because we cannot assume that agents are located in all ASs in the actual Internet, or agents can act with a large number of agents in all diagnosis phases. For example, the strategy determines a small number of agents that an agent should first access for diagnosis. When an agent starts performing detailed analysis, the agent may need information about other topologically suitable agents. This reorganization requires an on-demand search. Such location information on the BGP topology map is maintained by an agent organization management system called ARTISTE [11], which is an independent system of ENCORE-2. ARTISTE can search agents that match a given requirement, such as "Find agents that can relay messages and are located within 2 AS-hops from ASx".

3 Analysis Using Simulation Environment

We emphasize that the simulation model using the actual Internet topology is helpful to understand the routing behaviors. The distribution of connectivity values among ASs is an important index for determining diffusion of BGP routing information, but the connectivity is nothing more than an aspect of the network topology; other factors may also affect the behavior of routing information. Moreover, business relationships might have some statistical nature and might qualitatively affect the results. Ideally, various topological models that have special features should be checked. On the other hand, we cannot currently determine what is the dominant factors; probably they might affect each other. Instead, we introduced the actual Internet topology into the simulation with using simplified message distribution model, in order to understand the basic behavior of routing information under hijacking and preventing actions in relation to connectivity values of participant ASs.

3.1 BGP Topology

To analyze the behavior of BGP route distribution, we have constructed a BGP topology in a simulation environment constructed using Allegro Common Lisp / CLOS. In this simulation, each AS object has the AS-number, list of neighboring ASs and their preference for determining the best paths among possible alternative routes, and the BGP table. Each AS object also has communication queues for sending and receiving BGP information with BGP peers. The whole system is controlled using virtual time. An AS can send BGP messages to directly connected peers in one virtual time period. Hence, BGP messages are distributed to ASs in one AS-hop every virtual time period. This means that at least one advertised message reaches all ASs in n virtual time periods when the diameter of the BGP topology is n.

The BGP topology is a graph whose nodes are ASs and whose edges are the links among BGP peers. The topology is constructed using CAIDA's AS

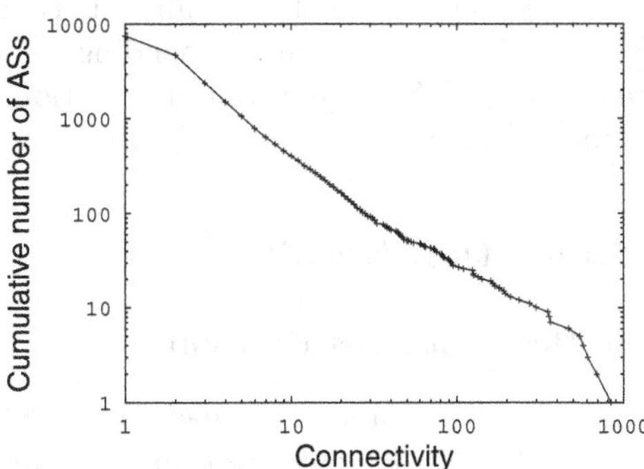

Fig. 2. Connectivity distribution used for simulation

adjacency data [12], which contains a list of BGP adjacency pairs. This Internet AS-level graph is computed daily from observed skitter measurements. The connectivity distribution of ASs in a generated topology is shown in Fig. 2. This topology was constructed using *skitter_as_graph.directed* data generated in March 2007. The total number of ASs appearing in this data set was 7490.

3.2 Simulation Conditions

In this simulation, we made some assumptions for simplicity and because of a lack of detailed information. In fact, these assumptions do not necessarily represent actual Internet management policies and might affect the accuracy of the simulation results, but we believe that the tendency of our simulation results is still meaningful.

- If an AS has a BGP connection, it can forward all received BGP information to connected peers. Its advertisement is not restricted by its local policy.
- The selection of the best path among possible BGP alternative entries is first determined by their AS-path lengths according to the selection rules determined by [1]. The `local_pref` values for implementing special local policies are not used.
- When length values are the same, preference values are used to determine the best path. These preference values correspond to IGP metric values and are randomly assigned to candidate peers when each AS object is created.

Although the virtual time cannot simulate real distribution time of each BGP update messages, the distance in AS-hop, which can be represented in the simulation using the virtual time, determines the best path selection. In other words, transitive states during messages are distributed vary with which messages arrive at what ASs faster, but the best-path selections after all messages have been completely distributed are determined only by the distance in AS-hop. Therefore, this simulation using virtual time is valid as far as analyzing how the best paths and/or hijacked BGP entries are statistically distributed. It is also important to know the distribution nature only depending on the actual network structure, because this nature is the base of the distribution that is affected by other factors, such as management policies of each AS.

4 Analysis of Simulation Results

4.1 Analysis of the Best Path Distribution

Fig. 3 shows how BGP-advertised prefixes spread from ASs that have typical connectivity values: 1, 30, 50, 103, 300, 468, 839 (max). It indicates that an advertisement from an AS that has a higher connectivity value (higher-connectivity AS) spreads faster. Because BGP messages are delivered in one AS-hop at one virtual time, this faster spread implies distribution in a smaller number of AS-hops.

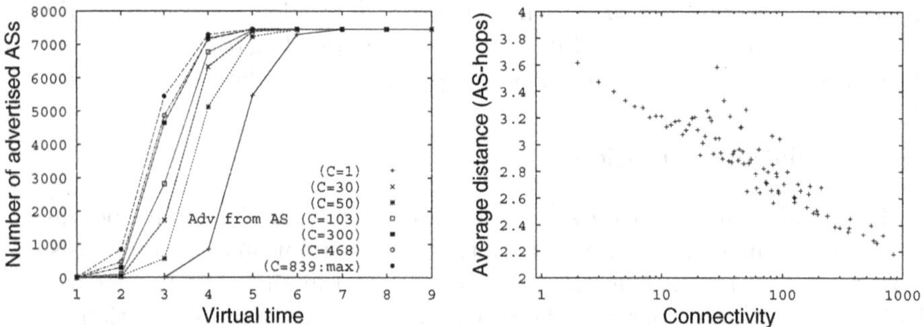

Fig. 3. Convergence of advertised messages and average distance to other ASs

Intuitively, the reason such distribution occurs is that such an AS can send advertisement messages to more ASs in the first virtual time period. This would result in rapid distribution in successive periods. Although this reason cannot explain all possible graph structures, it could be correct as far as the topology represents the actual Internet structure.

To confirm the situation described in the above paragraph, we analyzed the average distance of the best paths from each AS to all other ASs, as shown in Fig. 3. The average distance of an AS is defined as the average AS-hops of the best paths from the AS to other ASs. The average distance of connectivity value c, namely $D(c)$, is defined as the average distance of ASs whose connectivity value are c. These are formally defined as follows:

$$D(c) = average(\sum_{AS_i \in AS\_set(c)} distance(AS_i, AS_{j \neq i})),$$

where $AS\_set(c) \equiv \{AS_k | connectivity\_value(AS_k) = c\}$.

$$distance(AS_i, AS_j) \equiv as\_path\_length(best\_path\_from\_to(AS_i, AS_j))$$

where $best\_path\_from\_to(AS_i, AS_j)$ is the prefix advertised from AS_i and selected as the best path in AS_j, and $as\_path\_length()$ is the length of AS-path attribute of given prefix.

The result shows that an advertisement from a higher-connectivity AS arrives at other ASs via an AS path that has a shorter length than the path length of ASs that have lower connectivity values. Because the prefix of an AS-path that has a shorter length precedes that for a longer one, a higher-connectivity AS tends to have greater robustness against route hijacking. On the other hand, if a higher-connectivity AS also hijacks a route of another AS, this higher-connectivity AS tends to have a higher priority with respect to hijacking and could contaminate a larger number of ASs.

4.2 Analysis of Hijacked Route Distribution

To analyze this assumption in more detail, we checked the area that was contaminated under various conditions, where originating ASs with various connectivity values were hijacked by another AS with variable connectivity. The survival rate is defined as the ratio of the number of ASs where the correct route entry was selected as the best path divided by the total number of ASs. For example, the survival rate would be 0.5, when a hijacked prefix would be distributed and selected as the best path in a half of ASs.

Fig. 4 shows the rate of survivability against hijacking performed by other ASs. The survival rate was calculated by advertising a prefix from an AS (originating AS) and then injecting the same prefix from another AS (hijacking AS). The distribution of these advertisements stops in finite steps and this stable state can be checked by confirming there are no more messages to be sent.

The two horizontal axes show the connectivity of the originating and hijacking ASs. When multiple candidates that have the same connectivity values are possible, at most 10 ASs were randomly selected and experimented. For example, in the case where an AS with connectivity value of 10 were hijacked by another AS with connectivity value of 1, 10 ASs were selected in the experiment among ASs that have connectivity values of 10 as the origin AS. These ASs were checked in order. Then 10 ASs that had the connectivity value of 1 were randomly selected and did hijacking. Therefore, 10 * 10 cases were examined in this case. Fig. 4 also shows a more detailed view of some typical originating ASs having connectivity values of 1, 10, 50, 103, and 803 (max).

As predicted from the previous results concerning the average distance and connectivity, these results validate the assumption that a higher-connectivity AS has a higher survivability against route hijacking. In addition, if such a higher-connectivity AS performs route hijacking it can contaminate a larger number of ASs.

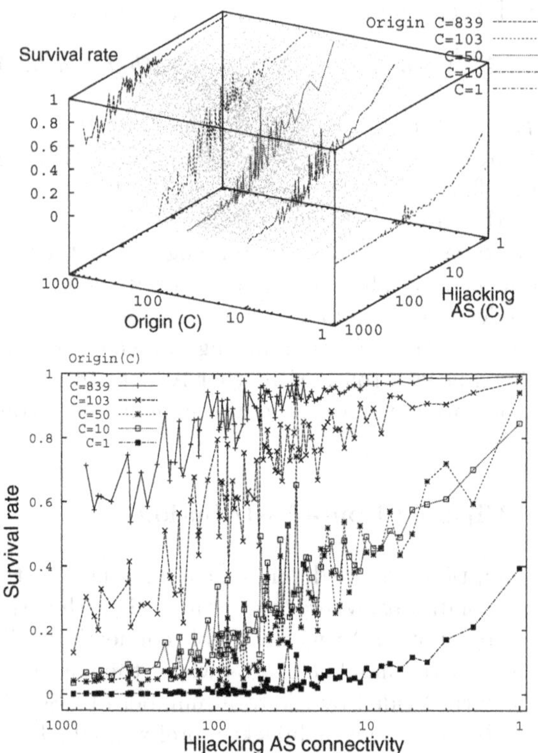

Fig. 4. Survival rate against hijacking

5 Discussion

Although fatal accidents or attacks like the 1997 failure that disturbed the entire Internet through the unintentional advertisement of full routes might not occur again, cases of small-scale or partial-area misadvertising have been observed several times in the past few years. Thus, continuous observation and diagnosis by an adequate number of sufficiently distributed agents are still needed. After detection and analysis, the operators of a misadvertising AS should be notified of their mistake as soon as possible.

On the other hand, some countermeasures should be applied to allow rapid though temporary recovery. One approach is to have ASs that detect hijacking perform filtering on this hijacked entry. Although an operator that is notified of this can set some filter commands on its routers by hand, which seems safer and more practical, some misconfiguration might occur or another error might be induced. Dynamic and on-demand filtering is possible by using methods like the VR architecture [6] and policy description of the agent, which cooperate with an agent that detects hijacking in the remote AS. However, more verification about the stability of such actions on the actual Internet is required.

In both cases, ideally all the ASs around the hijack-advertising AS should filter any advertisement involved, but in practice this is difficult to do. To analyze the effectiveness of this filtering action, we performed one more simulation that was similar to the hijacking analysis. The difference is that validity-check ASs were introduced. These validity-check ASs filter hijacked routes, which have an illegal originating AS number in their AS-path from the correct one.

The results are shown in Fig. 6. When multiple candidates had the same connectivity values, at most three ASs were randomly selected. The ASs that performed validity checking were selected from among high-connectivity ASs. These graphs are plotted for the cases where the number of checking ASs were 2^n: namely 1, 2, 4, ..., 512. An experiment where the connectivity value of an origin AS is set to 50 is also plotted as a 2-dimensional graph. In general, lager connectivity value of a hijacking AS indicates lower survival rate.

According to the previous analysis, the worst case is where a low-connectivity AS is hijacked by another AS having greater connectivity. All graphs in Fig. 6 also indicate lower survival rates in this case. On the other hand, these survival rates improve when the number of checking ASs is more than 8 or 16 even in the worse cases. This means that not all of the surrounding ASs are required for observation and counteraction. The deployment strategy where a certain number of agents in high-connectivity ASs check the route validity could be effective for detecting hijacked routes and preventing their distribution.

Another method is for an originating AS to advertise longer prefixes that include the hijacked address space. These longer prefixes can overwrite misadvertised routes because longer prefixes precede the currently advertised shorter ones according to the BGP best path selection rule. Although this dynamic injection of longer prefixes can also be executed using the VR architecture, the stability issue becomes more serious and more verification is required.

5.1 Application to MAS-Deployment Strategy

This subsection discusses the application of analyzed results to a MAS-based system using an example of ENCORE's actions. The BGP topology is assumed to be as shown in Fig. 5. In this example, an ENCORE agent R_1 in AS_1 asks a agent R_4 to observe BGP entries advertised from AS_1. R_1 notifies R_4 of the target IP prefixes in AS_1 and trap conditions. A typical trap condition is "Notify R_1 if the number of the originating AS, which should be AS_1, in any BGP entries concerning the target IP prefixes is changed or if any of these BGP entries disappear." According to analyzed results, an agent, which is requested to check the anomalous state such as hijacking on the actual Internet, should be deployed in an AS that has a higher connectivity value. The agent in such a AS can detect the change of AS-paths with higher possibility, because it has shorter average distance to all ASs as shown in Fig. 3

In this example, AS_1 advertises its IP-prefix-1 to AS_2. AS_2 advertises it to AS_3 and AS_5. Then, IP-prefix-1 is delivered and arrives at AS_4 via AS path (3, 2, 1). It also arrives at AS_7 via AS path (6, 5, 2, 1). If AS_7 misadvertises the same prefix IP-prefix-1 to AS_6, and AS_6 does not filter it, then this prefix from

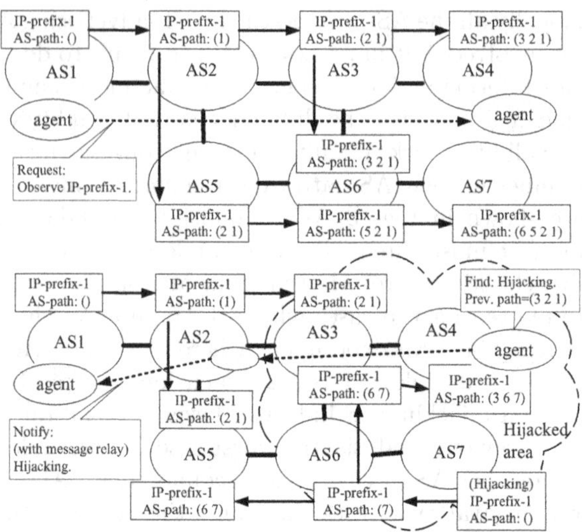

Fig. 5. Contamination by hijacked route delivery

AS_7 is selected as the best path for IP-prefix-1 because the BGP entry of IP-prefix-1 from AS_7 has a shorter AS path than the one from AS_1, so it has priority as the best path according to the default selection rules [1]. Similarly, the flow of routing information diffuses and contaminates routing tables in some other ASs. At this point, an agent deployed in a suitable AS could prevent successive contamination. The AS with a higher connectivity value on the actual Internet topology can be a candidate for checking the validity of BGP advertisement, as shown in Fig. 6.

In these contaminated ASs, packets destined for IP-prefix-1 are forwarded to AS_7. In this example, this misadvertised prefix also reaches AS_3 and AS_5. Here, AS_3 and AS_5 have two BGP entries concerning destination IP-prefix-1, and one is selected as the BGP best path. If the misadvertised route is selected in AS_3, then AS_4 is also contaminated. In this situation, R_1 is notified that the number of the originating AS, which should be AS_1, was changed to AS_7 in AS_4. Then, R_1 extracts possible hypotheses and starts verification.

A suspicious AS is found by comparing the AS paths of the two BGP entries acquired from the current BGP routing table in AS_4 and the previously existing one. The previous entries and update messages are recorded in buffers of the BGP-monitor in ENCORE and can be extracted. In this case, these are (3, 6, 7) and (3, 2, 1). The point where these two paths separate is AS_3. Therefore, R_1 first checks AS_3 and its neighboring AS, namely AS_2. Then, along with the current path, AS_6 and its neighboring AS_5 are checked. Therefore, R_1 can infer the contaminated area by repeatedly sending inquiries to investigation agents that are near the ASs located along the path where unintentional advertisement is detected. Although ENCORE uses the comparison between previous

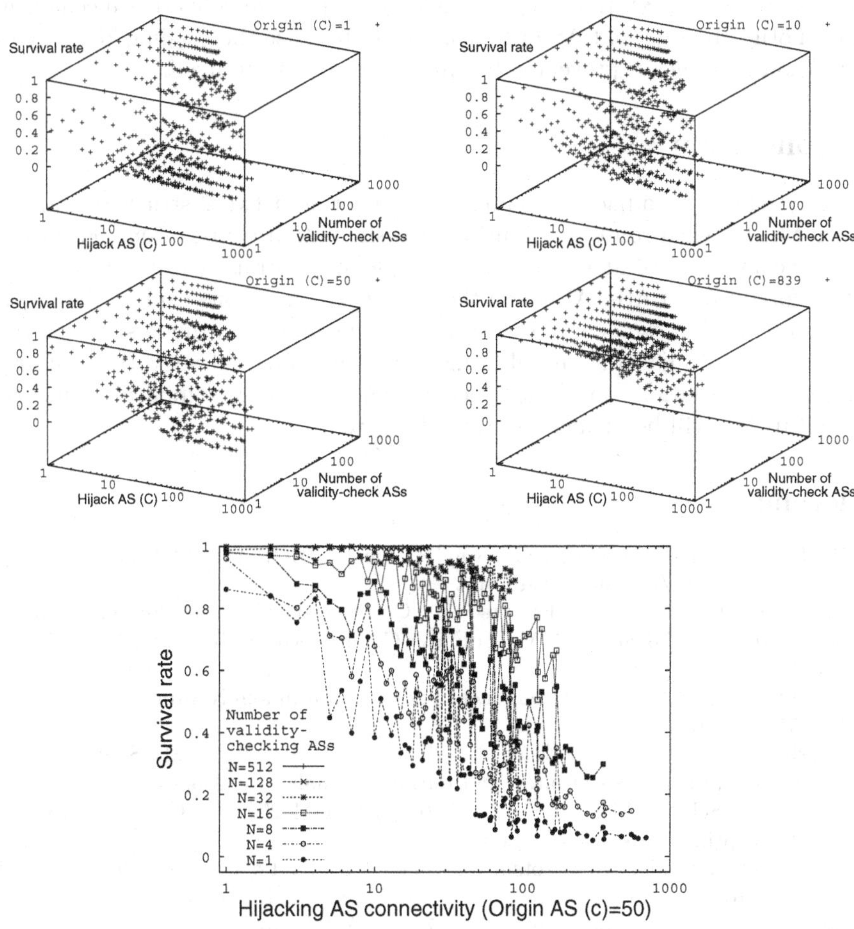

Fig. 6. Rate of correct routes: Effectiveness of validity-check ASs against route hijacking

and current BGP entries for detecting changes, other methods such as [8,9] can be incorporated as functions of MASs.

More specifically but importantly, the contaminated routing table prevents direct communication between AS_1 and AS_4 because packets in AS_3 destined for AS_1 are forwarded to AS_7. In this case, AS_1 and AS_4 can communicate indirectly by using a relay agent such as AS_2. The relaying agent also should be searched in ASs that have higher connectivity values, because the agents in such ASs have higher survival rates as shown in Fig. 4 and thus can relay messages among legal ASs. The analyzed results also show there are sufficient number of ASs that are not contaminated. Therefore, another strategy that makes agents near the legal origin relay messages is also possible. Note that a hijacked AS that does not have these cooperative actions can know only that some remote ASs such as AS_4 or AS_7 are inaccessible while some neighbor or nearby ASs are accessible. A system such as MyASN that has no cooperative message delivering mechanism cannot notify the hijacked AS of this situation because the sending/delivery of notification messages is prevented by this anomalous state itself.

6 Conclusion

We analyzed the AS adjacency topology of an actual network structure and the behavior of hijacked-route distribution in a simulation environment. From the results, we also devised an effective agent-deployment strategy using the connectivity information of ASs. Although these analyses were done using a simplified routing assumption, we believe the results are meaningful for understanding the basic behavior of hijacking and filtering actions in an actual network structure. More routing variation considering specific policies or more complicated deployment strategies will be included in the future work.

References

1. Rekhter, Y., Li, T.: A Border Gateway Protocol 4 (BGP-4). RFC1771 (1995)
2. Kern, E.: `http://nitrous.digex.net`
3. Faloutsos, M., Faloutsos, P., Faloutsos, C.: On Power-law Relationships of the Internet Topology. ACM SIGCOMM Computer Communication Review 29(4), 251–262 (1999)
4. Barabási, A.L., Albert, R.: Emergence of Scaling in Random Networks. Science 286, 509–512 (1999)
5. Akashi, O., Terauchi, A., Fukuda, K., Hirotsu, T., Maruyama, M., Sugawara, T.: Detection and Diagnosis of Inter-AS Routing Anomalies by Cooperative Intelligent Agents. In: Schönwälder, J., Serrat, J. (eds.) DSOM 2005. LNCS, vol. 3775, pp. 181–192. Springer, Heidelberg (2005)
6. Akashi, O., Fukuda, K., Hirotsu, T., Sugawara, T.: Policy-based BGP Control Architecture for Autonomous Routing Management. In: SIGCOMM workshops on Internet Netwrok Management, pp. 77–82. ACM Press, New York (September 2006)
7. RIPE, `http://www.ripe.net/`
8. Lad, M., Massey, D., Pei, D., Wu, Y., Zhang, B., Zhang, L.: PHAS: Prefix Hijack Alert System. In: Proc. of 15th USENIX Security Symposium, pp. 153–166 (2006)
9. Karlin, J., Forrest, S., Rexford, J.: Pretty Good BGP: Improving BGP by Cautiously Adopting Routes. In: Proc. of ICNP, pp. 290–299. IEEE Computer Society Press, Los Alamitos (2006)
10. Akashi, O., Sugawara, T., Murakami, K., Maruyama, M., Koyanagi, K.: Agent System for Inter-AS Routing Error Diagnosis. IEEE Internet Computing 6(3), 78–82 (2002)
11. Terauchi, A., Akashi, O., Maruyama, M., Fukuda, K., Sugawara, T., Hirotsu, T., Kurihara, S.: ARTISTE: An Agent Organization Management System for Multi-agent Systems. In: 8th Pacific Rim Int'l Workshop on Multi-Agents (PRIMA), (IFMAS). LNCS(LNAI), vol. 4078, pp. 245–259. Springer, Heidelberg (September 2005)
12. CAIDA: CAIDA's Macroscopic Topology AS Adjacencies, `http://www.caida.org/measurements/skitter/as_adjacencies.xml`

Analysis of BGP Origin AS Changes Among Brazil-Related Autonomous Systems

Shih Ming Tseng[1], Ke Zhang[1], S. Felix Wu[1], Kwan-Liu Ma[1], Soon Tee Teoh[2], and Xiaoliang Zhao[3]

[1] University of California, Davis, California, USA
{tsengs,zhangk1,wu,ma}@cs.ucdavis.edu
[2] San Jose State University, San Jose, California, USA
teoh@cs.sjsu.edu
[3] Juniper Networks Inc.
xzhao@juniper.net

Abstract. On the inter-domain Internet today, the address prefix origin in our BGP operations has become a major security concern. This critical problem can be stated simply as "Is the originating Autonomous System (AS) authorized to advertise the destination address prefix?" In the long term maybe we will be able to prevent this problem by applying proposed solutions such as SBGP[1] or SoBGP[2]. However, in practical network operations, it is critical to monitor and analyze all the BGP events potentially related to this BGP origin problem. In this paper, we have analyzed OASC (Origin Autonomous System Change) events, generated from the Oregon Route Views [4] archive, related to the Brazil BGP network. Our main focus is on how the Brazil BGP operation has been interacting with the rest of the Internet in the past five years. Also, we provide some possible explanations for OASC anomalies in Brazil.

1 Introduction

As the de facto inter-domain routing protocol, Border Gateway Protocol (BGP) [3] is responsible for discovery and maintenance of paths between distant ASes in the Internet. A BGP route lists a particular prefix (destination) and the path of ASes used to reach that prefix. The last AS in an AS path should be the origin of the BGP routes. We call that AS the *origin AS* of that prefix. A BGP prefix is normally announced by a single origin AS. However, in real BGP operations, for reasons such as multihoming [5], it is quite normal that some address prefixes are fully or partially originated by multiple ASes.

On the inter-domain Internet today, the address prefix origin in our BGP operations has become a major security concern. This critical problem can be stated simply as "Is the originating Autonomous System (AS) authorized to advertise the destination address prefix?" In the long term maybe we will be able to prevent this problem by applying proposed solutions such as SBGP or SoBGP However, in practical network operations, it is critical to monitor and analyze all the BGP events potentially related to this BGP origin problem. In

D. Medhi et al. (Eds.): IPOM 2007, LNCS 4786, pp. 49–60, 2007.

fact, we believe that, without a careful and clear understanding about the OASC events, it is very difficult to evaluate and justify any preventive solutions claiming to solve the origin AS problem in BGP.

While some of the observed OASC events are indeed due to normal operation, some others might very likely be related to failures, mis-configuration, or even, intentional malicious attacks. Previously, we have defined a set of OASC events [7,8], and developed a visualization tool to display the events graphically [6]. Ultimately, we would like to tell whether a particular OASC event is normal or abnormal, and furthermore, the explanation for our conclusion.

In this paper, we have analyzed the OASC events, generated from Route Views archive, related to Brazil BGP network. Our main focus is on how the Brazil BGP operation has been interacting with the rest of the Internet roughly in the past five plus years. Comparing to the whole Internet, Brazil ASes have produced slightly lower than average OASC events relative to its network size and address prefixes. Furthermore, we provide some possible explanations for certain OASC anomalous events in Brazil. More specifically, a few Internet-wide OASC storms identified in [6] have impacted Brazil prefixes significantly. Among other OASC anomalies we discovered, most of the Brazil- initiated OASC events in the summer/fall of 2003 were related to a very small number of ASes (such as AS6505 and AS7927), and, likely due to some legitimate reasons such as network re-configuration. Finally, we gave the definition of per-update OASC analysis, which shows very different results from the traditional per-day OASC case.

2 Origin Autonomous System Changes (OASC)

2.1 Definition of OASC

The origin AS number of a IP prefix P at time T_1 is denoted as $f(P,T_1)$ where function f is used to compute the origin AS number. Similarly, $f(P,T_2)$ represents the origin AS number of P at T_2. We define that there is an origin autonomous system change(OASC) event for prefix P between time T_1 and T_2 when $f(P,T_1)$ is not equal to $f(P,T_2)$. For instance, Multiple Origin Autonomous System (MOAS) change [7] is a special type of OASC events. A prefix P has MOAS property when P has two or more origin AS numbers. A MOAS change happens when P has MOAS property at time T_1 or T_2. (i.e. $f(P,T_1)$ or $f(P,T_2)$ contains two or more autonomous system numbers).

An OASC event implies the changes of BGP routing paths. Those changes are possibly caused by legitimate reconfiguration, unintentional mis-configuration or malicious attack. Regardless the causes, an increase in OASC events definitely implies an increase in network instability.

2.2 The Model of Snapshot Differentiation

A snapshot $S(T_1)$ is defined as a BGP routing table dump at time T_1. Assuming that there is a set of BGP update messages M which occurs between time T_1

and T_2. Hence, the snapshot S(T2) can be obtained after applying BGP update messages M to snapshot S(T1). The snapshot differentiation(SD) is defined as the difference between $S(T_1)$ and $S(T_2)$.

Per-Day-OASC. The per-day-OASC is defined as : T_1 and T_2 has the difference of 24 hours. Consequently, SD of $S(T_1)$ and $S(T_2)$ represents the changes of 24 hours. Also, M represents all of the update messages within the 24 hours duration. As mentioned in section 2.1, $f(P, T_1)$ is computed from $S(T_1)$ and $f(P, T_2)$ is computed from $S(T_2)$. Therefore, a per-day-OASC event of prefix P occurs if and only if $f(P, T_1)$ is not equal $f(P, T_2)$.

Per-Update-OASC. The per-update-OASC is defined as: BGP update messages M contains only one BGP update message. Hence, the snapshot S(T2) can be obtained after applying one BGP update messages M to snapshot S(T1). Similarly, for prefix P, a per-update-OASC event occurs if and only if $f(P, T_1)$ is not equal $f(P, T_2)$.

Obviously, per-update-OASC results the maximal level of detail in OASC analysis. However, per-update-OASC also results a huge computational overhead. For instance, it took us about 20 minutes to compute one snapshot based on the Per-Day-OASC model. But, it took us about one week of CPU time to compute all the Per-Update-OASC events of a 24 hours window. I.e., computing Per-Update-OASC is about 500 times more expensive.

Route Views BGP archive shows that (i) there are more then 262 million BGP update messages in total from all connected peers in March 2005 (nearly 100 update messages per second) (ii) there are more than 172,000 prefixes exist in the routing table dump of March 31, 2005. Due to the data complexity and limited computational resources, it might not be feasible to have per-update-OASC analysis by processing all BGP update messages. Instead, per-update-OASC method can be used as an auxiliary for per-day-OASC analysis when detail information is needed.

2.3 Types of OASC

Generally speaking, each OASC event contains five attributes [6]: (i) "$Time$" represents the date and time when an OASC occurs. (ii) "$Type$" represents the OASC event type (iii) "$Prefix$" represents the IP prefix in CIDR format (iv) "$OldAS$" represents a set of origin AS number before an OASC event. (v) "$NewAS$" represents a set of origin AS number after an OASC event.

There are mainly four categories of types of OASC events which are C_{type}, O_{type}, H_{type} and B_{type}. C_{type} event implies that the origin AS number of P is changed. O_{type} event implies a prefix P is announced and previously there is no route to P. B_{type} event implies that a more specific prefix P(e.g. /24 network) is announced with the same origin AS number as the origin AS number of less specific prefix P_{Big}(e.g. /16 network) where P_{Big} exists in routing table and P_{Big} is the smallest prefix which can embrace P. H_{type} event implies that a more specific prefix P is announced but the origin AS number is not the same as the origin AS number of less specific prefix P_{Big}.

In addition, there are up to four sub-types in each category. It comes out that there are twelve types of OASC event in total [6]. i.e. OS, OM, CSS, CSM, CMS, CMM, HSS, HSM, HMS, HMM, BSS and BMM. The abbreviation for each type of event is either in two or three letters. The first letter stands for the main category. The second and third letter which is either S (stands for *single*) or M (stands for *multiple*) means the number of origin AS number in $OldAS$ and $NewAs$. For examples, CSS stands for a C_{type} event and the number of origin AS number in $OldAs$ and $NewAS$ are both one.

3 OASC Analysis Results for Brazil-Related ASes

3.1 Analysis Goals

Our goals are (i) to design a series of OASC analysis methods which can be used to analyze BGP traffic systematically, (ii) to use the developed methods to give the best explanation of certain OASC events related to our target nation which is Brazil in this paper, and (iii) to find the relationship between OASC events and the AS topology of our target nation.

Our interests are to find the best explanation of the following questions: (i) Does Brazil have more OASC event than other countries ? (ii) Have Brazil networks been affected by major OASC storms in the past 5 years ? (iii) Have Brazil generated any OASC events which affected the rest of the world or only itself ?

3.2 Autonomous Systems to Country Mapping

Every autonomous system is represents by an AS number. We utilize the Whois server to map every AS number to its registered country. For example, AS6192 (University of California, Davis) is mapped to USA. From Route Views BGP archive and Whois servers, we are able to identify that there are a total number of 246 ASes registered to Brazil.

However, Whois servers only keep the most current information. In other words, we are not able to map some ASes due to lack of information. For example, an AS number X, which was registered in country Y in the past but X is currently unregistered and not assigned to any country. Under this kind of circumstances, it is impossible to map X correctly to country Y due to lack of history information. In this paper, the AS to country mapping process utilizes the data from Whois server on April 1, 2005.

3.3 Analysis Results

Route Views BGP archive provides the raw data of BGP traffic from January 1, 2000 to March 31, 2005. When processing BGP raw data, the per-day OASC analysis method is used to as primary tool to examine BGP traffic and per-update OASC analysis is secondary.

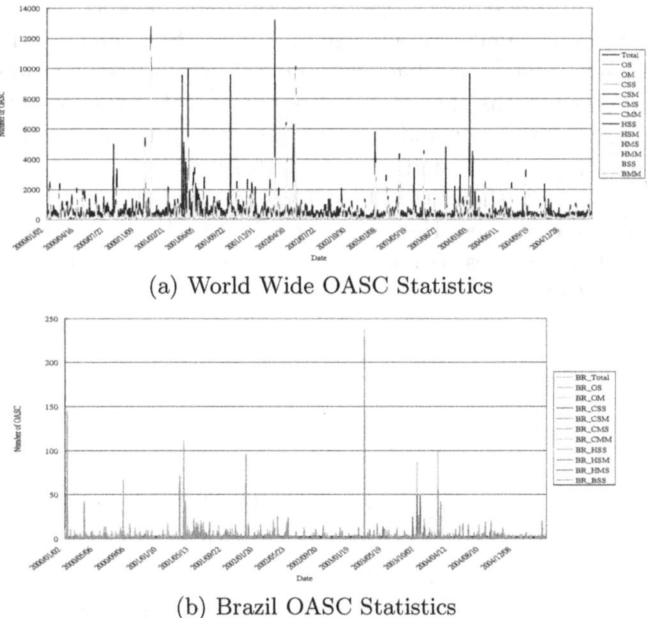

(a) World Wide OASC Statistics

(b) Brazil OASC Statistics

Fig. 1. OASC Statistics

The following abbreviations are defined for easy interpretation. (i) A Brazil Autonomous System(AS) is a autonomous system which is associated with Brazil in section 3.2. (ii) The Brazil-related OASC event is defined as a OASC event which involves at least one Brazil AS. (iii) A Brazil-originated OASC event is defined as an OASC event which is generated by a Brazil AS. In summary, a Brazil-originated OASC event is a Brazil-related OASC. But a Brazil-related OASC does not imply a Brazil-originated OASC event.

Compare Brazil-related OASC and Non-Brazil-related OASC. As a result from section 3.2, there are 246 Brazil ASes and 33788 non-Brazil ASes. The ratio of Brazil ASes to non-Brazil ASes is 0.728%. In addition, the total number of IP prefixes which have at least one Brazil AS as its origin AS is 1,435 and the nunber of all prefixes is 1,702,077 (BGP routing table of March 30, 2005). The ratio of Brazil prefixes to all prefixes is about 0.84%. In other words, on a per-AS basis, Brazil ASes own slightly more IP prefixes than other countries.

Fig. 1(a) and 1(b) show the number of OASC events in the past 5 years, for the whole Internet and the Brazil Internet, respectively. X-axis represents date and Y-axis represents the number of OASC events per day.

Fig. 2 shows the ratio of Brazil-related OASC to all OASC events where X-axis represents months from January 2000 to March 2005 and Y-axis represents the ratio. The per month average line indicates the ratio of Brazil-related OASC to all OASC events. Also, cumulative average line shows the average ratio from

January 2000 up to the end of a specific month. For instance, per month average is larger than accumulative average in August, September and October of 2003 which means that Brazil had more OASC events than long-term average during those three months. The figure also indicates that the long-term average (accumulative average) is around 0.5% to 0.6% which is smaller than the ratio of prefixes (0.84%) and the number of ASes (0.728%).

Overall, our analysis shows that Brazil has relatively fewer number of OASC events than other countries. However, according to Fig. 2, there were several statistical OASC anomalies. For instance, we can observe spikes on both April and October of 2003.

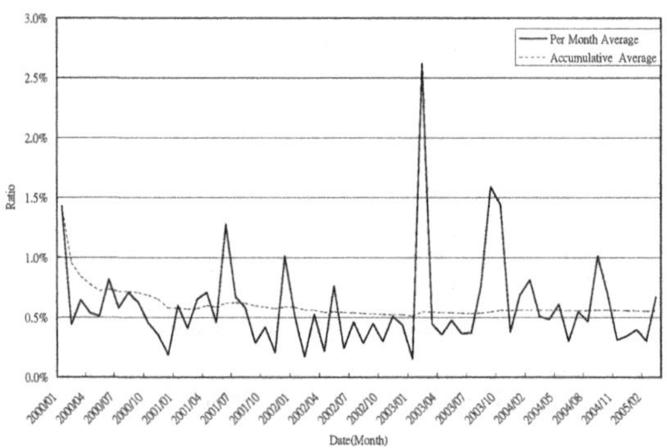

Fig. 2. Ratio of Brazil-related OASC to all OASC

Please note that some OASC events can not be recorded under certain circumstances. For instance, in Fig. 3, BGP router $R1$ makes some changes C on prefix P and the BGP update messages propagate to router $R2$ and $R3$. Based on the current status, $R2$ and $R3$ decide independently if C makes the best route to prefix P. If both $R2$ and $R3$ decide that C does not make the best route to P, the changes C will not be propagated. Hence, peer router PA and PB will not receive C such that C will not be recorded by Route Views. As a consequence, in order to record the changes C, a peer router must be placed near $R1$ to prevent C from being dropped by other BGP routers.

Currently, Route Views has more than 75 peer ASes. Hence, it is highly unlikely that some OASC events are not recorded by any of those 75 ASes. Currently, there is no observation point in Brazil. We can not exclude the possibility that some OASC events closed or within Brazil were not recorded due to aggregation by intermediate routers. In order to record an event before it is neutralized by aggregation, we suggest that a BGP router, which peers with Route Views, should be placed in Brazil.

Fig. 4 is generated by Route Views BGPlay service. The numbers represent ASes and the dash lines represent links. Fig. 4(a) shows the paths from 200.160.0.0/17 to all observation points and AS27699(Telecomunicacoes de Sao

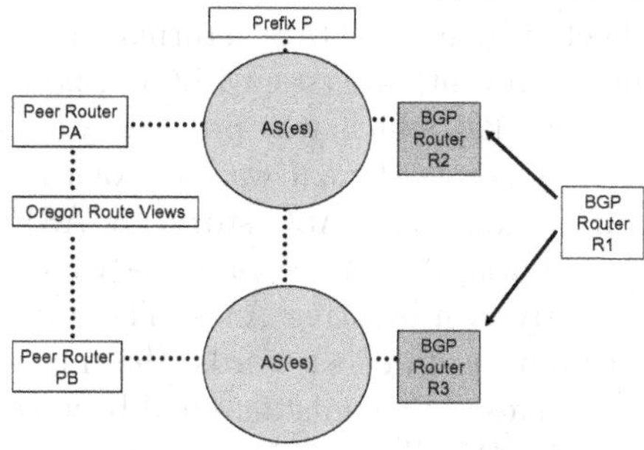

Fig. 3. Example of Non-recorded OASC

Paulo) is the origin AS number of 200.160.0.0/17.Similarly, Fig. 4(b) shows the paths from 57.74.160.0/19 and AS6505(Global One Communicaoes Ltda) is the origin AS number. On Apr 1, 2005, we find that there are 29 non-Brazil ASes which are connected to 46 Brazil ASes. Among those 46 Brazil ASes, AS7738 (Telecomunicacoes da Bahia), AS8167 (Telecomunicacoes de Santa Catarina) and AS13591 (MetroRED Telecom Services) have the largest number of connections to non-Brazil ASes. In addition, among those 29 non-Brazil ASes, 17 of them are in USA and the rest are: 2 in Argentina, 1 in Australia, 1 in Canada, 2 in Mexico, 1 in Italy, 1 in Spain, 1 in Switzerland, 1 in Germany, 1 in France and 1 in Uruguay. Please note that those 29 non-Brazil ASes have direct BGP peering sessions to 46 Brazil ASes. This fact implies that there are about 200 Brazil ASes which do not have direct connection across the country's border.

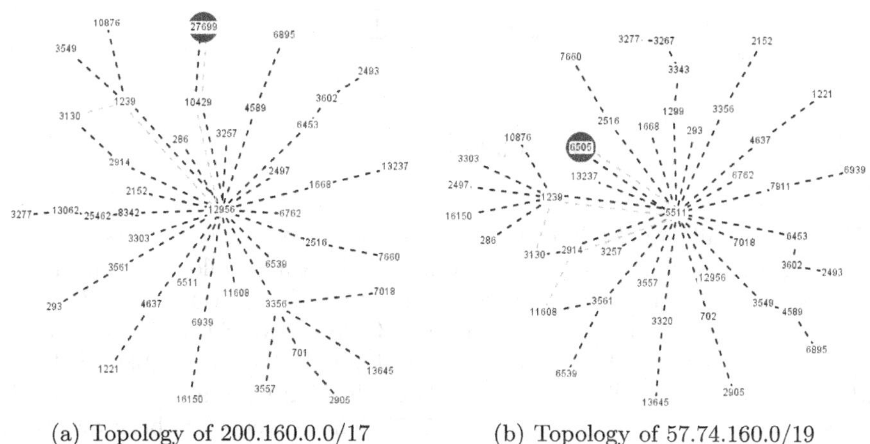

(a) Topology of 200.160.0.0/17 (b) Topology of 57.74.160.0/19

Fig. 4. AS Topology

Non-Brazil-Originated OASC Events Affect Brazil. As shown in Fig. 1(b), Brazil was affected by major OASC storms on August 2000 and April 2001. AS7777(Mail Abuse Prevention System LLC) announced a lot of smaller size prefixes (e.g. /30 networks) which was previously owned by other ASes. It results numerous of HSS events. Brazil was one of the victims. Afterward, Brazil was affected again by another OASC storm by AS15412(Flag Telecom). AS15412 announced that it was one of the origin AS number of numerous IP prefixes which was previously own by other ASes. The announcing caused CSM events in every affected prefixes. In other words, the multi-home property was added to those affected prefixes. Those mistakes had been corrected by dropping the paths announced by AS15412. Those fixing(dropping) procedures caused another OASC storm of CMS events. Overall, Brazil network was affected twice (CSM and CMS) by AS15412 in April 2001. Based on our observation, Brazil has not been affected by other major OASC storms after April 2001.

Brazil-Originated OASC Events. Generally, we can divide all Brazil-related OASC events into three sub-categories, *Non-BR to BR* , *BR to Non-BR* and *BR to BR*. Brazil-Originate OASC events are caused by Brazil-related AS initialed some changes in origin AS numbers.

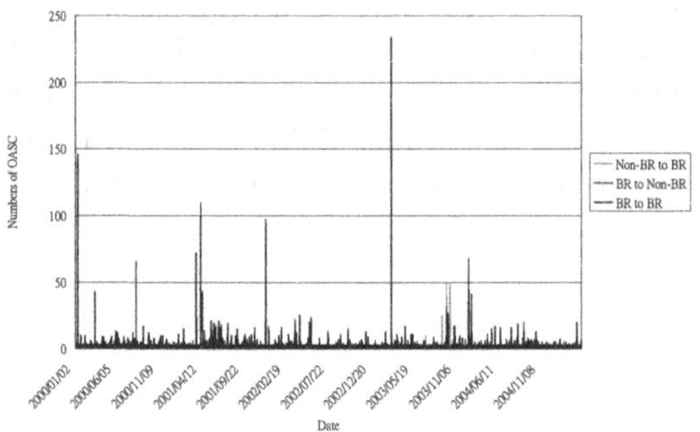

Fig. 5. Brazil-Related OASC Events

Fig. 5 shows all Brazil-related OASC events where X-axis represents date and Y-axis represents number of OASC event. We have the following conclusions by cross reference Fig. 1(b) and Fig. 5. (i) Most of the events in *BR to BR* sub-categories are due to the BSS ,CSM and CMS events. Since the BSS events only involve one single AS, the effects were limited. The noticeable CSM and CMS event were caused by AS15412. Also, there are rarely BR to Non-BR events except on August 14, 2000 by AS7777. Those events has been discussed in previously section. (ii) There are a cluster of Non-BR to BR events activities at the end of 2003. We have identified that those series of events are mostly CSS events related to AS6505. The data indicates that a set of IP prefixes previously own by Chile, Colombia and Venezuela ASes had been change to

AS6505. Furthermore, in the past 5 years, AS6505 had involved a total number of 242 OASC events and 147 of them are CSS events. We noticed that among all involved prefixes, most of the origin AS numbers still remain unchanged. Hence, we conclude that if those CSS events are legitimate, our best explanation is that was due to an large scale network aggregation from other countries to Brazil.

Fig. 6 shows the results of per-day OASC analysis and per-update analysis by using ELISHA visualization tool [6]. Fig. 6(b) highlights a small set of OASC events related to AS6505 and AS7927. A CSS (AS7927 to AS6505) event in per-day analysis could probably be more events in per-update analysis. A CSS event could possible be the consequence of two update messages(a withdraw and an announce) or one update message(an announce which replaces the old route). For example, prefix 206.228.36.0/24 was dropped from AS7927(Global One Venezuela) at 5:09am on October 12,2003 and added to AS6505 3 minutes later. The number of per-day and per-update OASC events is compared in Fig. 7. In conclusion, the number of per-update OASC event is greatly larger than per-day OASC and per-update OASC analysis gives more detail. As shown in Fig. 7, per-update analysis has a burst of events on October 9 , 2003 which is not shown by using per-day analysis. We leave the investigation of the burst of events for future research.

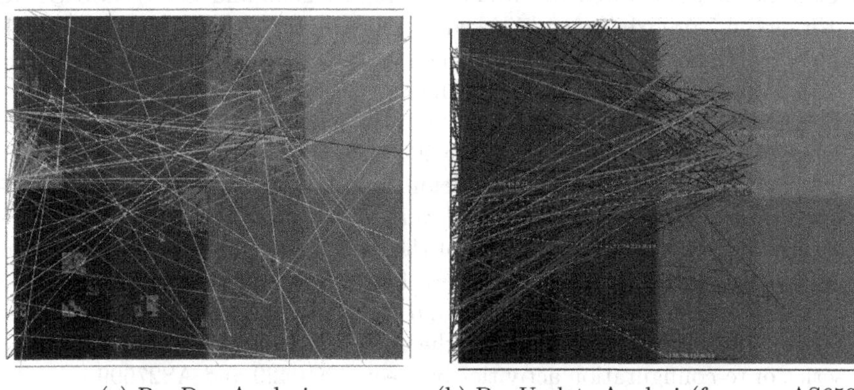

(a) Per-Day Analysis (b) Per-Update Analysis(focus on AS6505)

Fig. 6. Visualization of October 12, 2003 OASC events

Summary of Analysis Results. We summarize the dates which have more number of Brazil-related OASC events in Fig. 8. We have the following conclusions:

1. The majority of Brazil-related OASC events were due to numerous BSS events (i.e. "self punch a hole"). Those events were only involved with single particular Brazil AS and the origin AS number of most of the involved prefixes remains the same. Hence, we conjecture that those events might be legitimate re-configuration within Brazil.

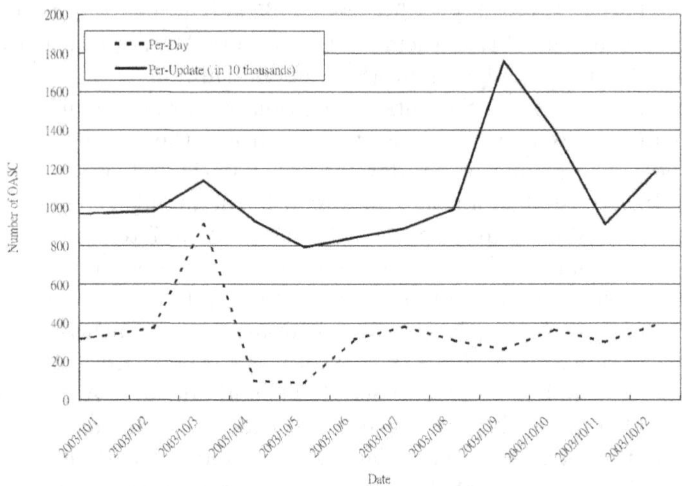

Fig. 7. Number of Per-Day OASC and Per-Update OASC

2. Brazil network was affected by the HSS OASC storm (i.e. "being punched a lot of holes") generated by AS7777 in August 2000 and CSM/CMS OASC storms (i.e. "prefixes being hijacked ") generated by AS15412 in April 2001.
3. The HSS events on October 16, 2003 might be legitimate because most origin AS number of involving prefixes remains the same.
4. The CSS (i.e. "ownership changed") events on August, September and October of 2003 were all involving AS6505 (Global One) in Brazil.
5. On October 10 2003, AS10429 (Telefonica Empresas) and AS27699 (Telecomunicacoes de Sao Paulo) in Brazil were involved in 27 HSS events. This is extremely unusual, while the rest of the world had only 287 events on that day, as Brazil had 27 events ($\sim$ 10%). Furthermore, we find a series of CSS and HSS events on October 27, 28 and 29, 2003. Among those events, most of the origin AS number remains the same. Hence, it might possibly be a series of re-configuration activities between AS10429 and AS27699.
6. Fig. 8(c) shows that AS13878 (Diveo do Brasil Telecomunicacoes Ltda), AS2715 (Fundacao de Amparo a Pesquisa do Estado de Sao Pau), AS1251 (Fundacao de Amparo a Pesquisa do Estado de Sao Pau), AS6505 and AS8167 (Telecomunicacoes de Santa Catarina) have involved a relatively larger number of OASC events among all Brazil ASes. We conclude that the number of OASC events related to an AS is not necessary proportional to the connectivity of the AS itself.
7. AS13878 has numerous of CSM and CMS events in the past 5 years. Among those AS13878-related OASC events, the origin ASes of those prefixes were changed back and forth frequently. There have been more than fifteen sets of prefixes having a similar pattern. For instance, the origin ASes of 200.9.219.0/24 (a Brazil prefix) were changed back and forth 16 times totally in three months (April, June, July of 2000), based on the Per-Day-OASC analysis.

Date	Total	OS	OM	CSS	CSM	CMS	CMM	HSS	HSM	HMS	BSS	Note
2000/01/12	146	5	0	0	0	0	0	0	0	0	141	AS2715
2000/03/16	43	1	0	0	0	0	0	2	0	0	40	AS10429
2000/08/14	68	0	0	0	0	1	0	66	0	0	1	AS7777
2001/03/21	72	0	0	0	0	0	0	0	0	0	72	AS14571
2001/04/06	112	0	0	0	104	0	1	1	0	0	6	AS15412
2001/04/10	42	4	0	0	0	37	0	0	0	0	1	AS15412
2001/04/12	43	2	0	0	0	41	0	0	0	0	0	AS15412
2001/12/07	97	0	0	0	0	0	0	0	0	0	97	AS17379
2002/04/05	26	2	0	0	0	0	0	0	0	0	24	AS13495
2003/02/26	237	3	0	0	0	0	0	3	0	0	231	AS1251
2003/08/24	26	0	0	24	1	1	0	0	0	0	0	AS7993(CL) to AS6505(BR)
2003/09/09	87	23	0	49	0	0	0	8	0	0	7	AS7984(CO) to AS6505(BR)
2003/10/06	27	0	0	0	0	0	0	27	0	0	0	AS27699
2003/10/12	50	0	0	49	0	0	1	0	0	0	0	AS7927(VE) to AS6505(BR)
2003/12/31	99	10	0	15	1	1	0	16	0	0	56	AS11097
2004/03/04	42	1	0	0	0	0	0	1	0	0	40	AS8167

(a) Large Number of Brazil OASC

Date	Percent	World_Total	BR_Total	BR_OS	BR_CSS	BR_CSM	BR_CMS	BR_CMM	BR_HSS	BR_BSS	Note
2000/03/16	9.21%	467	43	1	0	0	0	0	2	40	AS10429
2001/03/21	12.68%	568	72	0	0	0	0	0	0	72	AS14571
2001/06/02	8.91%	202	18	0	0	0	2	0	1	15	AS10733
2002/04/05	9.42%	276	26	2	0	0	0	0	0	24	AS13495
2003/02/26	8.06%	2940	237	3	0	0	0	0	3	231	AS1251
2003/08/24	19.85%	131	26	0	24	1	1	0	0	0	AS7993(CL) to AS6505(BR)
2003/10/06	8.60%	314	27	0	0	0	0	0	27	0	AS10429(BR) to AS27699(BR)
2003/10/12	12.85%	389	50	0	49	0	0	1	0	0	AS7927(VE) to AS6505(BR)

(b) High Ratio (Brazil to World)

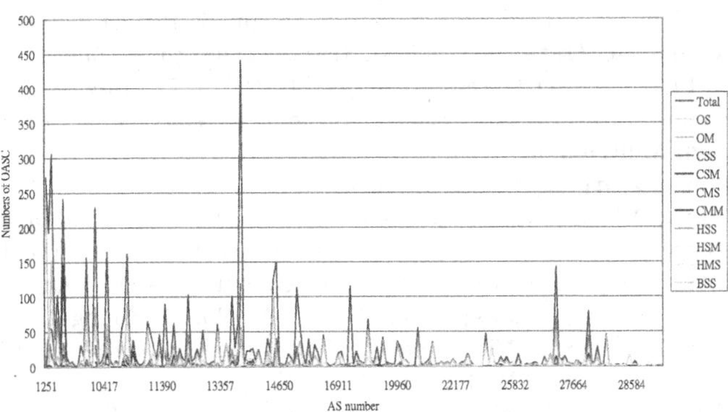

(c) Brazil OASC Statistics (Per AS number)

Fig. 8. Summary of Brazil OASC Anomaly

Unfortunately, we are not able to find any reasonable explanation of those CSM and CMS events. This phenomenon certainly causes negative effects on network stability.

4 Remarks

Origin AS Changes are global signals generated by the Internet BGP operations, intentionally or unintentionally. To analyze and diagnosis these causally related signals is the focus of our work here. We took the Brazil BGP network as the target domain and applied various techniques in trying to develop a systematic approach to decode the signals in depth.

As results, we have mixed news: good news, bad news, and unclear news. For the good news, we found that the Brazil BGP network is relatively stable compared to the rest of the Internet in OASC. And, even among the observed OASC events, we currently believe that most of them are legitimate. On the other hand, our bad news is that, not only Brazil has been hit by the OASC storms in the Internet, but also, in October 2003, a small number of Brazil-related ASes introduced unusually significant amount of OASC events into the Internet. Finally, due to the expensive CPU resources to compute the per-update OASC results, while we can observe a much greater details, we have very limited amount of per-update OASC information available. For instance, we observed a clear anomaly in 2000 for the prefix 200.0.219.0/24 and AS13877 on a per-day basis, but without a good explanation. We believe that the potential to explain OASC anomalies with the per-update OASC information is very significant.

References

1. Kent, S., Lynn, C., Seo, K.: Secure Border Gateway Protocol (S-BGP). IEEE JSAC Special Issue on Network Security (2000)
2. Ng, J.: Extensions to BGP to Support Secure Origin BGP (October 2002), `http://www.ietf.org/internet-drafts/draft-ng-sobgp-extensions-00.txt`
3. Rekhter, Y., Li, T.: Border Gateway Protocol 4. RFC 1771 (July 1995)
4. The Route Views Project, `http://www.antc.uoregon.edu/route-views/`
5. Smith, P.: Bgp Multihoming Techniques (2002), `http://www.nanog.org/mtg-0110/smith.html`
6. Teoh, S., Ma, K., Wu, S., Massey, D., Zhao, X., Pei, D., Wang, L., Zhang, L., Bush, R.: Visual-based Anomaly Detection for BGP Origin AS Change Events. In: 14th IFIP/IEEE Workshop on Distributed Systems: Operations and Management (2003)
7. Zhao, X., Pei, D., Wang, L., Massey, D., Mankin, A., Wu, S., Zhang, L.: An Analysis BGP Multiple Origin AS(MOAS) Conflicts. In: Proceedings of the ACM IMW 2001, Oct 2001, ACM Press, New York (October 2001)
8. Zhao, X., Pei, D., Wang, L., Massey, D., Mankin, A., Wu, S., Zhang, L.: Dection of Invalid Routing Announcement in the Internet. In: Proceedings of the IEEE DSN 2002, IEEE Computer Society Press, Los Alamitos (June 2002)

Securing a Path-Coupled NAT/Firewall Signaling Protocol

Sebastian Felis and Martin Stiemerling

NEC Europe Ltd., Network Laboratories,
Kurfuerstenanlage 36, 69115 Heidelberg, Germany
felis@netlab.nec.de, stiemerling@netlab.nec.de

Abstract. Dynamic configuration of IP Network Address Translators (NATs) and firewalls through application aware instances has been used within the Internet for quite some time. While current approaches, such as integrated application level gateway, are suitable for specific deployments only, the path-coupled signaling for NAT and firewall configuration seems to be a promising approach in a wide range of scenarios. Path-coupled signaling ensures that signaling messages and data flow are traveling the same route through the network and traversing the same NATs and firewalls. The path-coupled NAT/firewall signaling protocol is based on IETF's NSIS protocol suite. The NSIS-based NAT/firewall protocol specification is close to maturity and still needs a suitable and scalable security solution. This paper presents a framework to secure the NSIS-based path-coupled NAT/firewall signaling protocol across different administrative domains, based on zero-common knowledge security.

1 Introduction

Firewalls and Network Address Translators (NATs) are commonly used in the Internet for respectively protecting a network infrastructure and allowing private addressed hosts to communicate with other hosts using public IP addresses. Although the introduction of those devices brought indisputable benefits to networks interconnected to the Internet, these devices are severely impacting several critical applications, such as voice and video over IP, for today's Internet usage. These applications are breaking with a clear distinction between client/server roles, use unpredictable and rapidly changing port numbers for communication amongst each other. Potentially, their signaling flows and data flows use different paths through the network, involving different devices in one session. The peer-to-peer properties and the unpredictable port numbers are the most troublesome with respect to NAT/firewall. NATs and firewalls are mainly built for client/server applications, with at least one side having a static/fixed port number, such as HTTP with port 80. The devices can be pre-configured to allow traffic to particular destinations or fixed port numbers. Running peer-to-peer applications challenges the configuration of firewalls/NATs. Either it is a very restrictive configuration (well known in many Enterprise networks) blocking most traffic, or a very relaxed configuration (well known from home gateways on DSL lines or cable modems) allowing all

D. Medhi et al. (Eds.): IPOM 2007, LNCS 4786, pp. 61–72, 2007.

traffic. The first approach stops many applications from working, the second one leaves too much room for misuse and attacks.

1.1 NAT and Firewall Control

The best solution is choosing a path in the middle, running a configuration protocol from end hosts to NATs and firewalls, allowing end hosts to request a configuration for some time at the devices. There are related approaches allowing this already, such as MIDCOM [11], and integrated application level gateway (e.g., SIP ALG in device). However, all these approaches require either network topology knowledge or fail in more complex scenarios with several NATs or firewalls in parallel or sequential. A new approach is the path-coupled signaling for NAT/firewall traversal. Path-coupled signaling ensures that signaling messages and data flows are traveling the same route through the network and traversing the same NATs and firewalls. RSVP [10], for instance, is a path-coupled signaling protocol. [1] proposes a new path-coupled signaling protocol that dynamically configures on-path packet filtering firewalls and NATs, so that subsequent data flows can traverse them without any problem. This protocol allows a controlled dynamic configuration of NATs and firewalls, independent of the network topology and independent of the type of device to be encountered. Each time a path-coupled signaling enabled firewall or NAT is met, the device is configured either by opening firewall pinholes or installing NAT bindings. This protocol is currently developed within the framework of Internet Engineering Task Force's (IETF) Next Steps in Signaling (NSIS) working group [13] and is called NATFW NSLP. This working group works on a more generic IP signaling protocol beyond RSVP. The work presented in here aims at a extentsion of the NSIS NATFW NSLP standard in the near future, as the current specifications are close to their final publication.

1.2 Security Threats

Using a firewall/NAT external configuration ability raises a fundamentally question about security. Firewalls and NATs are used for security reasons to protect networks and hosts from a wide range of attacks. A device external configuration ability, apart from remote access through the network administrator, looks counterproductive on the first glimpse, if not properly secured. Typically, network administrators are the only instance that is allowed to change the configuration of these devices. In some cases trusted integrated application level gateways or third party off-box devices (see [11]) are allowed to perform limited changes to the configuration at runtime. Gaining control over the signaling protocol would result in control of the firewalls and NATs running this protocol. Therefore, the protection of the NAT/firewall signaling protocol is a key aspect for a successful deployment of it. So far the current protocol specification has not yet mitigated all security threats to the protocol. In the following sections, we propose a security solution for the NATFW NSLP, without requiring deployment of a public key infrastructure for the purpose of authenticating hosts or users.

Section 2 introduces briefly the NSIS protocol stack and the NAT/firewall signaling application part. Section 3 evaluates specific security threats and describes the security requirements. The solution is presented in Section 4 and related work is presented in Section 5. The paper concludes with Section 6 and gives a short outlook on future work.

2 Path-Coupled NAT/Firewall Signaling Protocol

Path-coupled IP signaling ensures signaling messages and data flow are traveling the same route through the network, meaning that both are taking the same links and visiting the same routers within the network. A comprehensive definition of path-coupled is available in [4]. Figure 1 shows the effect of path-coupled signaling in the Internet. The networks of host A and of host B are separated from the public Internet by firewalls and/or NATs, depicted as boxes labeled with MB for middlebox. The route from A to B is different as from B to A, in the case of routing asymmetry. With a non path-coupled signaling protocol, data and signaling may take independent and non correlated routes through the network. The figure shows that signaling and data messages from A to B are taking exactly the same route and traversing the same MBs due to the path-coupled property of the signaling protocol. Vice versa, this holds true for messages from B to A.

NSIS proposes such a protocol, based on a two-layer architecture for the next generation IP signaling protocol [2]. The common layer is the NSIS Transport Layer Protocol (NTLP, [4]) and is responsible for the path-coupled signaling message transport between neighboring NTLP nodes. NTLP nodes are routers implementing the NSIS signaling protocol suite. The signaling intelligence itself is embedded into the NSIS signaling Layer Protocol (NSLP). The NSLPs implement the signaling application that is responsible for interacting with the resources to be configured. These resources can be, but are by no means limited to, quality of service [5], NAT/firewall [3] traversal, and traffic meter configuration [12]. The single NSLPs are only implemented on devices supporting the particular configuration of resources and NSIS signaling messages are only processed by nodes implementing the NSLP signaled for. In the remainder of this text the term NSLP refers to the NATFW NSLP if not specified otherwise.

The entities of the NSLP protocol are data sender (DS), data responder (DR), NSIS initiator (NI), NSIS responder (NR), and NSIS forwarders (NFs). NFs are located at the firewall or NAT, the NI is typically located at the DS, and the NR is typically located at the DR. The protocol supports scenarios where DS and NI or DR and NR are not co-located. These scenarios are not described herein due to space limitations but are considered in the security solution. The remaining document assumes that DS/NR and DR/NR are co- located. Figure 1 shows the entities and a possible location within the network. Firewall/NATs are typically

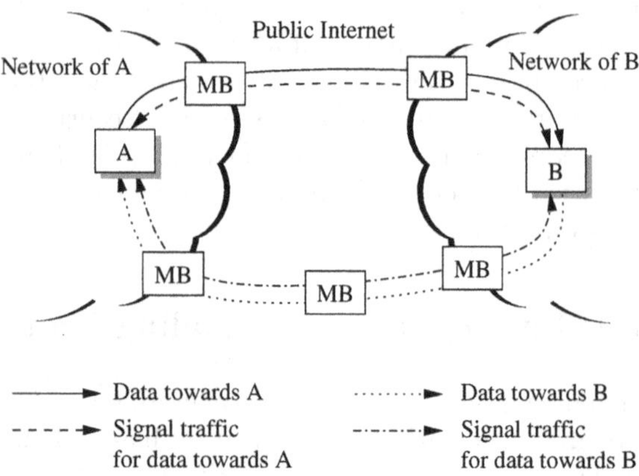

Fig. 1. Signaling and data flows for asymmetric up- and downstream between two communication parties

located at the edge of the network and less in the core of the network. These devices protect single network, such as network of A and B in the figure.

An NI(A) intending to send data to NR(B) signals first with the NSLP towards the address of NR. Each intermediate NF implementing the NSLP intercepts the signaling message, processes it, and finally forwards it again towards NR. The message is terminated at the NR and is responded to by NR. The reply travels hop-by-hop backwards to NI, visiting so all nodes as before (this is similar to RSVP's RESERVE and PATH message exchange). NFs configure their firewall and NAT engines according to the signaling request received and the subsequent data flow can now be sent from NI to NR, traversing all devices unhindered. This signaling session actually consists of a NTLP signaling exchange and a NSLP signaling exchange. See Figure 2 for an overview about the whole NSIS stack. The transport between neighboring NTLP nodes uses UDP for the first message exchange and can be switched to a different transport level protocol. TCP, TCP/TLS, and IPsec are currently defined as further transport protocol for the NTLP. Figure 2 shows different combinations of transport protocol and layers in NSIS. The signaling session is started with an NTLP message exchange towards B's address. If there is any NSLP node, such as MB in Figure 1, that node responses on the NTLP layer. After completing the NTLP three way handshake, both nodes exchange the NSLP data. This three way handshake must be performed between all neighboring nodes. This process continues until the signaling message arrives at B. B returns a reply message back to A on the prior established signaling path (as described above).

The installed firewall and NAT configuration and the protocol state itself are called a signaling session; the handle to identify the session along the whole path is called session ID. The signaling session is maintained through a soft-state, meaning that if not refreshed by new signaling messages the session will be removed automatically. The NI is in charge of initiating, refreshing, and

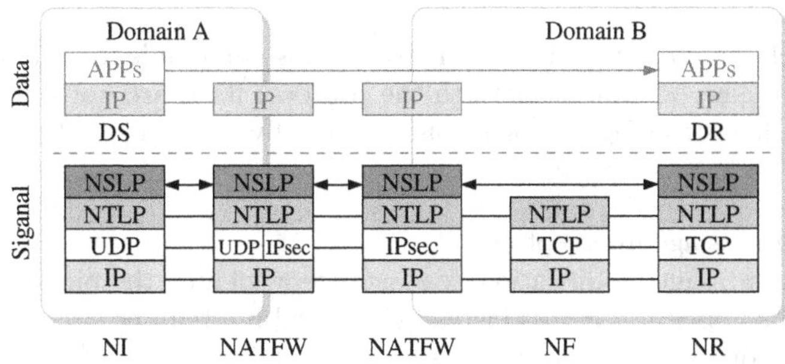

Fig. 2. Stack of NSIS signaling and application data. Signaling stack and data stack are independent but located at the same path. The signaling messages are bidirectional, whereas the application data flow is unidirectional from NI to NR.

terminating sessions. The NFs can generate asynchronous messages and terminate the session at any time.

3 Security Environment

Firewalls, and to some extend NATs, are very sensitive with respect to security and opening these devices to NAT/firewall signaling protocols increases their vulnerability to attacks significantly. Attacking the signaling protocol can give access to the device itself and to a service provided by the device not intend to be used. Firewalls and NATs are typically located at the network edges (see Figure 1) and controlled by the administrative domain they are belonging to. Entities not belonging to this administrative domain do not have any control over these devices. Firewalls/NATs within a domain are assumed to be aware about other firewalls/NATs in their domain, since they are all under the same control of one administration. This domain membership can be used to provide information about neighboring devices, such as IP addresses, hosts certificates, and pre-configured shared secrets. On the other hand, an administrative domain does not know about properties of another administrative domain. These properties are ingress points, firewalls/NATs, used security infrastructure (if any at all), security policies, etc. It cannot be assumed that there is a global public key infrastructure. However, administrative domains may easily setup their own local public key infrastructure. In certain scenarios, such as interconnecting telecommunication networks, it can be assumed that both network know each other's peering point, by either common certificates and/or IP addresses. On the other hand, for smaller unmanaged networks it can be assumed that there is no knowledge about each other and that there is no knowledge about the internal devices within the network, meaning zero-common (without pre-shared) knowledge.

3.1 Threats to the Protocol

An unprotected NAT/firewall signaling protocol is vulnerable to on-path and off-path attacks. The possible attacks to the protocol itself are not different from attacks to other signaling protocols but potentially cause a significantly higher damage. Possible attacks are:

- signaling message interception,
- signaling message modification, for instance, widening the pinhole specification and so open full access instead of desired limited one,
- replaying of messages to refresh signaling session state although it should have timed out,
- injecting messages to teardown state, for instance,
- message flooding, causing denial of service,
- signaling session hijacking,
- man in the middle attacks and
- wiretapping of signaling messages.

All threats are mitigated by our security solution except the message flooding and to some extend man in the middle attacks. The message flooding mitigation is part of the initial NTLP design already specified in the protocol specification. The session hijacking is taken here as a security threat showcase. All nodes involved in the signaling session must be able to validate incoming requests if they have been issued by the NI, since the NI must be the only one being able to create, maintain, update and delete sessions. The NI is called session owner. Requests for a session cannot only be received through an already established path (from where the initial message was received) but also from any other path. The requirement of the NSLP being able to react to network route changes and to support node mobility is the reasoning of this. If the session ownership validation is not given, any node, either on or off-path, can try to delete a session or try to divert a session by using a state update message. For instance, an off-path attacker located somewhere in the Internet could send signaling messages to one of the NFs, indicating a path change, and so changing the complete signaling path plus firewall/NAT configuration, rendering it useless for the actual NI.

The full list of protocol security threats is described in detail in Section 7.2 of [3].

3.2 Requirements

The protocol must be protected from the initial message exchange, establishing the session state, till the deletion of the state. The goal is to achieve a relationship between the neighboring NTLP/NSLP nodes so that they and only they are able to send and receive messages for a particular signaling session. Other, non session nodes, may send messages towards session members, but their messages must be discarded. The keys used to encrypt the signaling messages between the nodes should be negotiated during the NTLP session up.

Authentication and Authorization. The protocol must support the transport of authentication and authorization information (AA). Intermediate NSLP nodes can use this information to decide whether the initiator is known and allowed to request a specific configuration. The policies for checking the allowed configuration can be stored device-locally or on any policy control server remotely. This policy checking architecture is out of scope of this paper.

Sender Invariance. Despite the fact of being able to transport authentication information the protocol needs a proof of session ownership mechanism without necessarily requiring an NI's identity to be known. This ownership identification must be valid across changes of location (node mobility, change of IP addresses) and changes of authentication information (for instance, changing from one authentication system to another). Furthermore, Section 3 describes the fact that there is no global key infrastructure, rendering an authentication from one side of the network useless on the other side of the network. It is required to have an independent session owner identifier, to recognize the NI without any knowledge about his real identity.

Hop-by-Hop Security. NSLP nodes may be located far away in terms of IP hops from each other, crossing untrusted networks, or links, possibly revealing NAT/firewall configuration to other parties. This requires a hop-by-hop encryption of the signaling protocol, so that the configuration data, authentication information, etc is secured between nodes. The goal of the hop-by-hop security is to secure the whole signaling path from the NI, via the NFs, towards the NR, so that only these nodes can receive and send messages within this session. Other hosts, not belonging to the signaling path, either on-path or off-path, must not be able to tamper with messages.

Next Node Identification. Another aspect is the next NSLP/NTLP node discovery, even though NIs can be identified, transmission between neighboring nodes can be encrypted, messages are sent into the network and nodes can claim to be a NAT/firewall running NSIS. It should be possible, if applicable, to identify trusted NSLP nodes with the protocol. NSLP nodes should be able to authenticate mutually if desired. Otherwise, they should be at least able to establish an encrypted connection even though they are unknown to each other.

4 Proposed Solution

This section describes our proposed solutions for securing the path-coupled NAT/firewall signaling protocol. For each requirement a solution is given.

For next next neighbor discovery, the NTLP sends *QUERY* messages (UDP with an IP router alert option set) towards the address of NR to discover the next NSLP node on the path. Nodes along the path implementing the NTLP and the requested NSLP intercept these *QUERY* messages and reply with a *QUERY RESPONSE* (see Figure 3). The handshake is completed with a *QUERY CONFIRM* and thus the signaling session is established. Now, the NSLP data is transmitted to the next node. The protocol assumes proper nodes (nodes not being attackers) to be the only nodes replying to this and has currently no measure to double-check this. Any node is able to claiming being a NSLP node even though it is an attacker. For the further discussions we distinguish between two cases: managed infrastructure and unmanaged infrastructure (as described in Section 3). There is no way in unmanaged environments to check the next nodes claim to be a NSLP node. This holds true for complete networks when they are communicating with NSIS and they do not know each other in advance. The neighboring nodes simply can trust each other. In managed networks neighboring NSLP nodes are likely to know each other by several means. One common possibility is using certificates. All NSLP nodes within the managed infrastructure will be assigned a certificate by the local network administration. If a node claims to be a NSLP node, this node must, for instance, present its certificate or a derived signature within the message exchange. Otherwise, a node receiving a *QUERY RESPONSE* as reply on its *QUERY* discards the reply. Vice versa, nodes receiving *QUERY* may require a certificate to be presented as well. The remaining text assumes that nodes running this next node discovery always accept the identification presented by the other node. The next section describes the full handshake between neighboring nodes and extends it to a fast and secure setup.

4.1 Fast Hop-by-Hop Encryption Setup

The NTLP specification proposes the use of TCP with SSL or IPsec to secure subsequent message exchanges. TCP has been chosen as reliable transport protocol for long running communications carrying multiple NSLP session, thus reusing the TCP connection. While this reusing holds true in cases where signaling messages travel (at least) partially the same route it is not applicable in cases where the data routes are too diverse, meaning that every data flow takes a different route. The usage of TCP or IPsec is quite expensive if not used for multiple sessions at the same time. For every new session it is required to run the complete setup exchange if not an already established connection can be reused. In the TCP/TLS case 1.5 round trip times (RTT) for TCP and 2 RTT for TLS will follow the UDP exchange. In the IPsec case, assuming the usage of IKE with certificates, it takes 3 RTT for phase 1 and 1.5 RTT for phase 2. These times count always between neighboring hops and sum up along the signaling path since multiple nodes are involved. Taking into account that this NSLP signaling runs concurrent with the application level signaling it is desirable to keep setup times as small as possible. A good example for such an application level protocol is SIP, with call setup times around a second. Delaying this call setup due to very long NSLP signaling times on setup is undesirable.

We propose to extend the UDP-based discovery mode by introducing a Diffie-Hellman (DH) key exchange during the discovery phase. The result is an encrypted connection right after the *QUERY CONFIRM* message, after 1.5 RTT only. Combined with the authentication approach in managed infrastructures of Section 4 connections are authenticated and encrypted directly after the three-way UDP handshake and do not require any additional TCP/TSL or IPsec setup. This solution does not preclude scenarios where TCP/TLS or IPsec is desired.

Each NTLP discovers the next NTLP node with a 3-way handshake (as described earlier) and includes its authentication data. Middlebox A of the neighboring node in Figure 3 sends a NTLP *QUERY* towards the address of NR. This message is intercepted by middlebox B. This middlebox replies with a *QUERY RESPONSE* to A. This *QUERY RESPONSE* contains the allowed authentication types and the DH challenge. B must be able to specify a list of authentication types it is able to handle and willing to offer. The authentication types are, but not limited to, a X.509 certificate, a Kerberos ticket, a SSH key, or user/password pair. If A is able to provide the requested authentication it sends a *QUERY CONFIRM* to B, containing its authentication, its DH response, the public key of NI (see Section 4.2) and the NSLP data. All this data must be already encrypted with the shared secret key obtained from the DH exchange. Any further data exchange from A to B and vice versa must be encrypted.

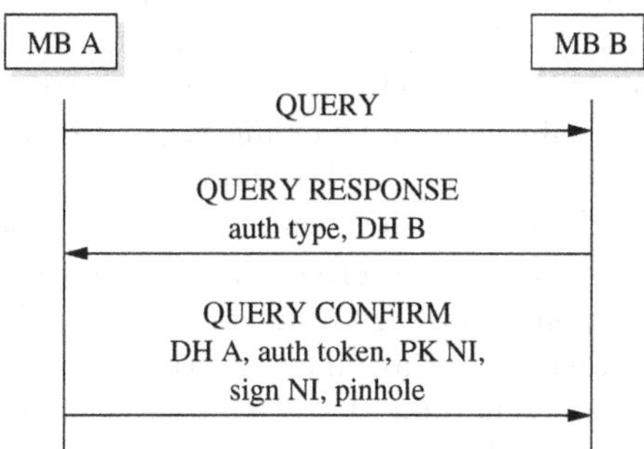

Fig. 3. NATFW NSLP authorization Handshake within the 3-way discovery handshake of NTLP

4.2 Sender Invariance and Replay Attacks

The NI must provide an authentication along the signaling path as required in Section 3.2 independent of location and locally used identification. [8] describes a way of authentication of a network communication where the actual identity of the source is not important but it is important that successive messages in the communication come from the same source. The use of a temporarily generated public/private key pair for each new signaling session to be established is proposed.

The NI must generate a new public/private key pair for every new session and must include a handle and a signature into each message to be sent. The handle is required for the intermediate nodes to correlate messages with stored state and is the session ID (SID). The session ID is derived by hashing the public key and thus it correlates to the public key. A message sent by the NI must carry this session ID, a message sequence number (MSN), and the digital signature done with the public key over the message. The MSN is increased by one for each message sent and serves together with the signature against replay attacks.

The public key must be distributed along the signal path to the different NSLP nodes so that they are able to verify the signature of the messages. A possible way is the usage of a key distribution system but this assumption collides with the unavailability of a global public key infrastructure. However, it is required to distribute the public key along the path and the proposed way is to include it in one of the signaling messages from the NI towards to the NR. Taking this into account, the NTLP signaling exchange is extended in this way (see Figure 3): the first *QUERY* message sent from A to B at the NTLP layer carries just the SID and MSN, but no signature since the public key is not known anyway. B stores the SID and sends the response *QUERY RESPONSE* to A. The *QUERY CONFIRM* message additionally contains the public key and is signed. B stores this public key and every subsequent message arriving at B must be signed with this key.

This procedure must be repeated between all neighboring nodes along the signaling path. All nodes can identify the correctness of received messages after the first message exchange is complete. It is important to identify the NI along the complete signaling chain without knowing its real identity, as described in Section 3.2. However, there is an attack window in the setup phase since the first message is unprotected and can be intercepted and modified in a man in the middle attack. There are no countermeasures against this attack when both nodes do not share common knowledge in advance. The presented solution does not preclude the additional use of an infrastructure based identity on parts of the network. It is possible to combine the PBK approach with a locally available network identifier, such as a X.509 certificate-based signature. Hosts can sign the signaling message and the local network uses this information to verify the correctness of the messages and if hosts are allowed to request the configuration. However, it is likely that networks on the other side of the signaling path are not able to re-use this signature.

5 Related Work

The work presented here spans a wide range of topics, such as protocol design host identification, etc. It is impossible to reference to all possible related areas in this paper and thus we chose to discuss only very closely related areas of different topics handled in this paper.

Authentication systems: There are many related techniques in the area of host and network authentication, such as X.509 certificates or Kerberos system, Naming them all in the framework of the path-coupled NAT/firewall signaling protocol clearly exceeds the paper's scope. The final specification will consider these techniques and provide a proper container for their transport.

RSVP security: RSVP defines its own security mechanisms in [10] that mainly relies on prior common knowledge between RSVP nodes. This certainly does not satisfy the requirements for path-coupled signaling protocols with less prior or zero-common knowledge. The same holds true for the proposed extension of RSVP as firewall signaling approach in [14]. The mechanisms are basically the same as for standard RSVP.

Ad-hoc networking: The problem of little or zero-common knowledge about the other communication partners is quite similar to the situation in mobile ad-hoc networks. These networks consists out of mobile nodes that have never seen each other before and have the need to establish a secure communication. [9] proposes a "zero-common" knowledge authentication protocol for this scenario. The path-coupled NAT/firewall signaling protocol runs partially in the same domain, networks that never have had any connection will need to cooperate in a secure way without any global security infrastructure and zero-common knowledge.

Group security: Section 4.1 can be referred to as establishing a group security between all nodes of a signaling session. All nodes of the session should be able to read, process, and modify messages, but no other. Our approach is a group security one based on hop-by-hop keys but not on a group key. In a group key system all nodes could read all communication of all other nodes. Possible group key models fitting to path-coupled protocols are the grouped Diffie-Hellman GDH.1 described in [6] or the use of keyed hash chains based on [7]. The GDH.1 approach can be used to secure the signaling path with a single end-to-end NSLP message exchange. This approach has a major drawback for the NSLP environment. Each time the data path changes (some NFs drop out, some NFs join in), the NI must initiate a complete re-keying cycle end-to-end to establish a new group key. The hash chain suffers from the same problem of re-keying in the event of path changes and requires additionally a mechanism for re-starting the chain after all elements have been used. With our hop-by-hop approach the establishment of keys is a local problem between two neighboring nodes and advantageous compared to re-keying, re-starting, and implementation overhead.

6 Conclusions

This contribution presents a security solution for the NSIS-based path-coupled NAT/firewall signaling protocol. The solution mitigates all threats, except message flooding and to some extend man in the middle attacks. The message flooding is expected to be covered by the NTLP. A key aspect of the solution is the zero-common knowledge between NSLP nodes or networks. Nodes or networks not known in advance can setup a secure signaling session and so mitigate attacks. However, an attack window is still present in this approach and man in the middle attacks cannot be mitigated. Scenarios with pre-configured knowledge, such as pre-installed shared secret, can prevent man in the middle attacks completely.

The described solution has been implemented on a Linux-based NATFW NSLP prototype using IPTables and is considered as work in progress. It has been tested for semantically correctness in video on demand and voice over IP scenarios. Future work will concentrate on large scale scalability and performance tests. Furthermore, the details about hashing functions, public/private key methods, and to be used DH method will be evaluated in future work.

Acknowledgment

The authors would like to thank Juergen Quittek, Dirk Westhoff, Hannes Tschofenig, Cedric Aoun and Hannes Hartenstein for their valuable comments.

References

1. Martin, M., Brunner, M., Stiemerling, M., Fessi, A.: Path-coupled signaling for NAT/Firewall traversal. In: IEEE HPSR 2005, Kong Kong (May 2005)
2. Hancock, R., Karagiannis, G., Loughney, J., van de Bosch, S.: Next Steps in Signaling: Framework. In: RFC 4080 (June 2005)
3. Stiemerling, M., Tschofenig, H., Aoun, C., Davies, E.: NAT/Firewall NSIS Signaling Layer Protocol (NSLP). Internet Draft (work in progress, 2007) (draft-ietf-nsis-nslp-natfw-14.txt)
4. Schulzrinne, H., Hancock, R.: GIST: General Internet Signaling Transport. Internet Draft (work in progress, 2007) (draft-ietf-nsis-ntlp-13.txt)
5. Manner, J., Karagiannis, G., McDonald, A.: NSLP for Quality-of-Service signaling. Internet Draft (work in progress, 2007) (draft-ietf-nsis-qos-nslp-13.txt)
6. Steiner, M., Tsudik, G., Waidner, M.: Diffie-Hellman Key Distribution Extended to Group Communication. In: Proceedings 3rd ACM Conference on Computer and Communications Security (1996)
7. Lamport, L.: Password authentication with insecure communication. Communications of the ACM 24(11), 770–772 (1981)
8. Bradner, S., Mankin, A., Schiller, J.I.: A Framework for Purpose-Built Keys (PBK), Internet Draft (January, 2003) (draft-bradner-pbk-frame-06.txt)
9. Weimerskirch, A., Westhoff, D.: Zero-Common Knowly Authentication for Pervasive Networks. In: Matsui, M., Zuccherato, R.J. (eds.) SAC 2003. LNCS, vol. 3006, pp. 73–87. Springer, Heidelberg (2004)
10. Braden, B., Zhang, L., Berson, S., Herzog, S., Jamin, S.: Resource ReSerVation Protocol (RSVP). RFC 2746, Version 1 Functional Specification. RFC 2746 (September 1997)
11. Srisuresh, P., Kuthan, J., Rosenberg, J., Molitor, A., Rayhan, A.: Middlebox communication architecture and framework. RFC 3303 (August 2002)
12. Fessi, A., Kappler, C., Fan, C., Dressler, F., Klenk, A.: Framework for Metering NSLP. Internet Draft (October 24, 2005)
13. IETF NSIS working group (June 2007), `http://www.ietf.org/html.charters/nsis-charter.html`
14. Roedig, U., Goertz, M., Karsten, M., Steinmetz, R.: RSVP as Firewall signaling Protocol. In: Proceedings of the 6th IEEE Symposium on Computers and Communications, Hammamet, Tunisia, IEEE Computer Society Press, Los Alamitos (July 2001)

DiffServ PBAC Design with Optimization Method

Ricardo Nabhen[1,2], Edgard Jamhour[1], Manoel Penna[1], Mauro Fonseca[1], and Guy Pujolle[2]

[1] Pontifícia Universidade Católica do Paraná, Brazil
[2] LIP6,Université Pierre et Marie Curie, France (Paris VI)
{rcnabhen,jamhour,penna,mauro.fonseca}@ppgia.pucpr.br, guy.pujolle@lip6.fr

Abstract. Determining the maximum amount of traffic that can be admitted in a DiffServ network is a difficult task. Considering a realistic traffic scenario, the relationship between the traffic load and the queue length distribution of a DiffServ node is very difficult to model. This paper demonstrates how a non-liner programming (NLP) algorithm can be employed to determine the maximum load that can be accepted by a DiffServ node without deriving an analytical model. The NLP algorithm is used to "train" a parameter based admission controller (PBAC) using a specifically designed traffic profile. After training the PBAC for a specific network and specific statistical QoS guarantees, it can be used to provide these guarantees to distinct offered traffic loads. This method was evaluated in a sample scenario where (aggregated on-off) VoIP traffic and (self-similar) data traffic compete for the network resources.

1 Introduction

In a DiffServ network [1], the traffic can be classified into distinct aggregated classes, according to its performance requirements. For example, VoIP traffic can be assigned to a class designed to provide low delay and data traffic can be assigned to a class designed to provide low packet loss. In order to determine the maximum amount of traffic that can be admitted in a DiffServ network, it is necessary to determine the relationship between the traffic load and the queue length distribution of a DiffServ node. This relationship is strongly dependent on the incoming traffic profile and the queuing discipline adopted by the DiffServ node. Because of the incertitude on the traffic nature, characterize a model for the queue length distribution is a difficult task.

In the literature, it is possible to find several proposals for designing admission control (AC) algorithms for DiffServ networks. These AC algorithms can be divided into two large categories: PBAC (parameter-based access control) and MBAC (measured-based access control). The PBAC approach defines the AC parameters based only the theoretical traffic behavior, the queuing disciplines and the network capacity. Usually, the PBAC proposals assume simplifications on the traffic behavior in order to provide an asymptotic bound for the queue tail probability. The MBAC approach, by the other hand, measures the real network traffic in order to dynamically adjust the AC parameters. In fact, MBAC approaches enable to design controllers that are more robust with respect to the accuracy of the traffic model. However, MBAC controllers

D. Medhi et al. (Eds.): IPOM 2007, LNCS 4786, pp. 73–84, 2007.

are also dependent on some sort of parameter tuning, and a setting that gives excellent performance under one scenario, may give a very pessimistic or too optimistic performance in another scenario [2-3].

Even though several techniques have been proposed for designing AC algorithms for providing QoS guarantees to the aggregated traffic, designing AC algorithms for providing QoS guarantees to individual flows is an issue far less addressed. As pointed by some authors, under certain conditions, the delay, jitter and packet loss level experienced by individual flows can significantly diverge with respect to the performance of the aggregated traffic [4-5]. Consequently, even when the aggregated traffic is properly dimensioned, the individual flows may not satisfy QoS requirements such as delay and packet loss. Providing individual flow guarantees poses additional difficulties on the DiffServ AC design.

This paper addresses this problem using an alternative approach. We employ a non-liner programming (NLP) algorithm to determine the maximum load that can be accepted by a DiffServ node without deriving an analytical model. The NLP algorithm is used to "train" a parameter based admission controller (PBAC) by using a specifically designed traffic profile. After training the PBAC for a specific network and specific statistical QoS guarantees, it can be used provide these guarantees to distinct offered traffic loads. Because the NLP is a multi-dimensional optimization process, this approach can be applied to, virtually, any AC technique (PBAC or MBAC) with one or more tuning parameters. The difficult in using this method, however, relies on choosing the adequate training scenario. To illustrate the proposed approach, a sample scenario where (aggregated on-off) VoIP traffic and (self-similar) data traffic compete for the network resources was simulated and evaluated.

The remaining of this paper is organized as follows. Section 2 reviews some related works concerning the design of AC algorithms. Section 3 presents the rationale that motivates the development of our work and describes the strategy adopted to generate the traffic profiles for training the controller. Section 4 presents the optimization method used for finding the optimum AC parameters. Section 5 presents and discusses some numerical results. Finally, section 6 concludes this work and points to future developments.

2 Related Works

A common approach for designing a DiffServ AC algorithm is to determine the maximum amount of traffic that can be aggregated without leading to an excessive buffer overflow. A single-link analysis of several statistical PBAC algorithms that follows this approach is provided by Knightly and Shroff in [6]. Also, surveys comparing the performance of both, PBAC and MBAC algorithms have been presented in [2-3]. The works discussed on these surveys assume that satisfying the QoS requirements for the aggregated traffic is sufficient to satisfy the QoS requirements for the individual flows. Some works have shown, however, that the end-to-end delay and the packet loss level of individual flows can substantially fluctuate around the average aggregate performance.

A study presented by Siripongwutikorn and Banerjee [4] evaluated the per flow delay performance with respect to the traffic aggregates. The authors consider a scenario with a single node, where heterogeneous flows are aggregated into a single class. Distinct queuing disciplines have been considered, such as FIFO, static priority, waiting time priority and weighed fair queuing. The authors observed, by simulation, that the traffic heterogeneity, the load condition and the scheduling discipline affect the per-flow delay performance. Notably, the simulation results indicate that it may not be able to achieve delay guarantees for some individual flows based solely on the class delay guarantees when the flows are heterogeneous in a high load condition.

The work presented by Xu and Guerin [5] explored the differences that can exist between individual and aggregate loss guarantees in an environment that enforces guarantees only at the aggregate level. The work develops analytical models that enable to estimate the individual loss probabilities in such conditions. A bufferless single hop scenario with distinct traffic sources have been considered: ON-OFF, constant bit rate periodic and real traffic video sources. The authors points that in order to avoid significant deviations across individual and aggregate loss probabilities, one should avoid multiplexing flows with significantly different rates into aggregates with a small number of sources. They also observed that the per-flow deviation decreases with the number of aggregated sources. The number of sources required to reduce the deviation across flows is significantly higher when the ON-OFF sources have rather different peak and mean rate.

Spliter and Lee [7] have presented an optimization method for configuring a node level-CAC controller for a bufferless statistical multiplexer. The problem assumes a single class of traffic, which must provide QoS expressed in terms of packet-loss constraints. The arriving call-requests are assumed to follow a Poisson distribution and the call durations are assumed to be general. The flow corresponding to each call is modeled as an ON-OFF process. The traffic during the ON period is assumed to be constant. The idea of the CAC controller was to minimize the call blocking probability subject to the packet loss constraints. Because the authors assume a buffeless approach, the optimization process could be modeled in terms of a linear programming problem. The packet loss ratio constraint was imposed to the aggregated traffic instead of to individual flows. The authors present then a CAC policy where the decision about the admission of a call is based only on the number of calls in progress. In spite of using an optimization approach, this work differs from our proposal in several points. First, we assume a multi-class scheduler with buffer capabilities. Second, our AC algorithm is designed to provide per-flow guarantees. Finally, our method is based on non-linear programming approach and does not require a mathematical model to describe the constraints imposed to the problem, which enables to employ the method to complex traffic profiles.

3 Scenario and Proposal

In order to evaluate the method proposed in this paper, we have considered a sample scenario where (aggregated on-off) VoIP traffic and (self-similar) data traffic compete for the network resources. The VoIP traffic is controlled by an AC algorithm which limits the number of simultaneous active VoIP flows. We consider homogeneous VoIP flows aggregated into a single EF class. The self-similar data traffic is not controlled by the AC algorithm. Instead, it is submitted to a leaky bucket classifier, which determines which packets will receive an assured forwarded (AF) treatment (i.e., packets within the limits of an accorded service rate) and which packets will be treated as best effort (BE). The AC strategy must provide statistical guarantees that each VoIP flow will respect a percentile limit imposed on the end-to-end delay and packet loss performance. This scheme must also provide delay and packet loss performance guarantees for the AF data traffic.

In this paper, we are interested in designing a simple PBAC algorithm, which must be capable of answering the following question: what is the maximum number of simultaneous VoIP flows that can be served without violating the performance guarantees?

As discussed in section 3, the per-flow problem is not extensively addressed in the literature. Also, most attempts of deriving a mathematical model for the per-flow behavior assume that the life-cycle of all VoIP flows is identical, i.e., all flows start in the beginning of the evaluation and terminates simultaneously. In real world, however, the life cycle of VoIP flows is variable. This variability can be modeled by assuming two additional parameters: TBF (time between flows) and AFD (average flow duration) [8]. In our study, both, AFD and TBF represent the average value of exponentially distributed random variables. Our work shows that the effect of considering flows with distinct life cycles introduces significant deviations among the individual flows percentile performance, requiring a "per flow" tuning of the AC method in order to respect the QoS requirements.

In order to train our PBAC controller we have selected a variable VoIP offered load template which follows a "peaked call arrival pattern" [8], as defined in Fig. 1. This approach is necessary in order to design a PBAC controller which is robust with respect to the fluctuations of the offered load. According to the figure, the evaluation time is divided into ten identical periods. Each period corresponds to a distinct offered load, which is determined by adjusting the TBF parameter. The AFD parameter depends on the average user behavior, and it is kept constant among all periods. The offered load is computed in terms of an estimation of the blocking probability, by using the Erlang-B functions. In this case, the equivalent number of VoIP lines for a given link capacity is "virtually" defined by the AC method, which imposes a limit to the number of simultaneously active VoIP connections.

The data traffic used to train the PBAC algorithm follows the self-similar model presented by Norros, the Fractional Brownian Motion [9]. According to this model, the accumulated aggregate arrival process is represented as follows:

$$A_t = mt + \sqrt{am}Z_t, \quad t \in (-\infty, \infty) \tag{1}$$

Where Z_t is a normalized fractional Browniam motion (fBm), m is the mean input rate, a is a variance coefficient and H is the Hurst parameter of Z_t. In this work, we have adopted a = 275 kbit.sec and H = 0.76, as used by Norros in the sample simulation presented in the same seminal paper. The discussion of the values of these parameters is out of the scope of this work. The value of the m parameter is adjusted according to the target average occupation rate of the link capacity. In this case, the variable arrival pattern is generated by the noise represented by the $\sqrt{am}Z_t$ component.

For each Δt window we compute the arrival process and generate a packet series $A_{\Delta t}$ assuming a MTU size (1500 bytes) and a minimum packet size of 46 bytes. The variable Z_t was implemented using the Hosking Method [10]. The packet series is then submitted to a leaky bucket metering scheme, controlled by two parameters: the bucket size and the transmit rate. The $A_{\Delta t}$ packets within the bucket limits are tagged as AF PHB, and the non-conforming $A_{\Delta t}$ packets (i.e., packets that could not be inserted into the bucket) are tagged as BE. In this paper, we have adopted $\Delta t = 10$ ms.

The scheduling discipline of the DiffServ node is a priority queuing mechanism with three non-preemptive finite size queues: EF, AF and BE. In this scheme, a queue is served only when the higher priority queues have no packets waiting to be served.

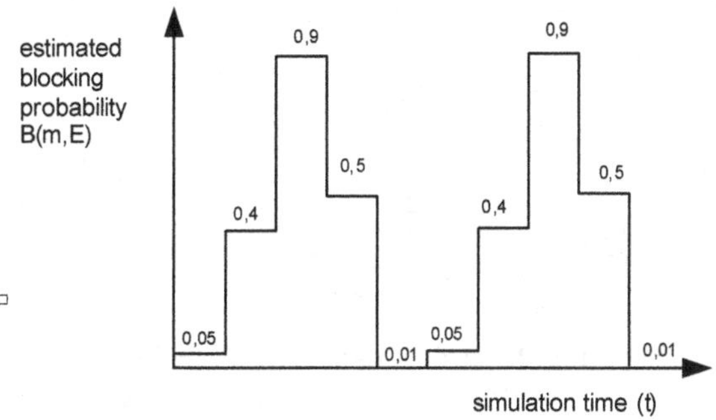

Fig. 1. Variable offered load template

Figure 2 shows the results of a simulation that illustrates how the individual flows performance can diverge with respect to the aggregate performance. In this scenario a load of connections has been offered to a PBAC algorithm that imposes a limit of 60 simultaneously active flows. Please, observe that this number of flows was not determined using our optimization method. It was chosen intentionally high in order to illustrate how difficult it is to predict the behavior of individual flows by taken into account only the aggregated traffic behavior.

Each VoIP connection was modeled as an ON-OFF source. As in many VoIP modeling works (see [11], for example), we have assumed that both ON-OFF intervals are exponentially distributed. The exponential averages were defined as 0.4s for the ON period an 0.6s for the OFF period. The total packet size was 200 bytes, already included the payload and the protocol headers. During the ON period, the arrival interval between packets is 20 ms, which represents a peak rate of 80 kbps and an average rate of 32 Kbps for each individual flow. We have limited the VoIP (EF) queue size to 50ms, i.e., packets that exceed the 50ms delay are dropped. The AF queue size was limited to 250ms. The simulation presented in Fig 2 has considered the following additional input parameters: link capacity: 2 Mbps and AFD: 210s. The m parameter in equation (1) was adjusted to represent 50% of the link capacity. The leaky bucket used to classify the data packets was set to a size of 2 MTUs and a transmit rate of 50% of the link capacity.

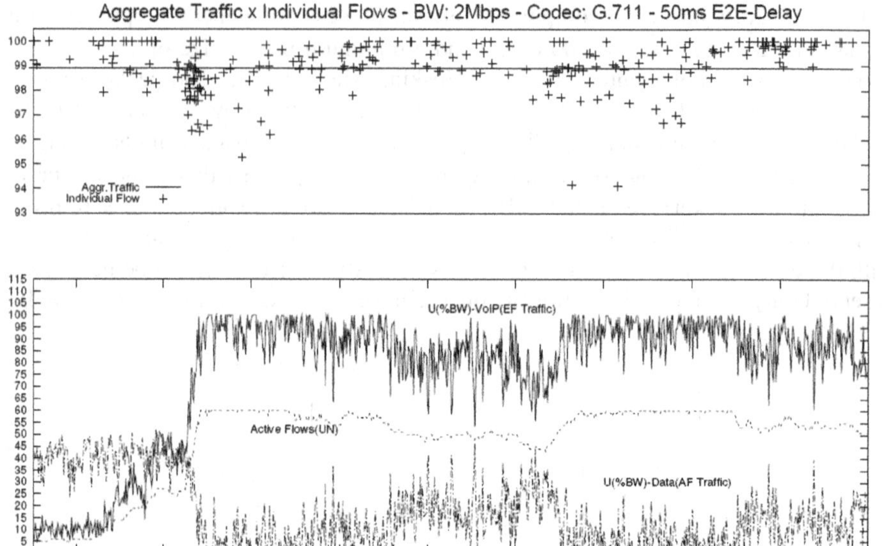

Fig. 2. Per-flow variability illustration

One observes in the superior part of figure that, approximately, 99% of the VoIP packets have satisfied the delay bounds. However, the "per-flow" percentile of packets that satisfied the delay bounds has significantly diverged with respect to the aggregate performance (each cross in the figure represents the performance of an individual flow plotted at the instant when the flow terminates). In this scenario, the drop level experienced by the AF class was about 56% (not shown in this figure).

It is clear that in order to design an AC algorithm it is necessary to take simultaneously into account both, the per-flow VoIP and the AF traffic requirements. Because in our scenario the AF traffic is limited by a leaky bucket, it is possible to find conservative estimates for its bandwidth requirements (see [12], for example). For the VoIP traffic, however, finding an analytical estimate is a hard task, as most of the proposals found in the literature target the aggregate performance (see [6], for a survey). We conclude that, when per-flow guarantees are required, even the simple scenario depicted in this section is too hard to be analytically treated. That motivates the proposal in this paper of designing a PBAC algorithm using an optimization approach.

The PBAC algorithm addressed in this paper is quite simple. We assume that the PBAC algorithm is capable to determine the exact instant when each VoIP flow starts and terminates. Therefore, the only tuning parameter of this controller is the maximum number of simultaneously active flows.

4 Optimization Method

The optimization method adopted in our work is called "The Flexible Polyhedron (FP) Method" [13, 14]. It is a "search method" for solving non-linear programming

(NLP) problems without constraints. The motivation for choosing the FP method is that it does not require mathematical derivations of the cost function, which enables to combine this method with computational simulations in order to solve problems related to complex network scenarios. The main idea of the FP method is to determine the search direction based on the best solutions found in the previous interactions. A geometric figure called polyhedron is formed by a set of vertices, where each vertex correspond to a solution **x**. The number of vertices is equivalent to the number of the variables of the optimization problem plus one. During each interaction, a new solution is computed by applying arithmetic operations on the Polyhedron vertices, and the worst vertex is replaced. Considering a successive set of interactions, the tendency is the polyhedron to adjust itself around the optimum solution and reduce its dimension (distance among the vertices) until a convergence criterion is achieved. The FP method can generate infeasible solutions during the search process, raising a difficulty for treating constrained optimization problems. This drawback can be avoided by making the cost function to strongly penalize infeasible solutions.

Suppose $\mathbf{x} \in \Re^n$ as a vector formed by the variables that one desires to optimize. An optimization problem can be defined in terms of a cost function $f_c(\mathbf{x})$: $\Re^n \rightarrow \Re$. For each element **x** belonging to the solution set, $f_c(\mathbf{x})$ represents the "level of rejection" of the corresponding solution. In order to be valid a solution **x** must be feasible, i.e., it must satisfy a set of constraints related to the nature of the problem being solved. Let **X** be the space of feasible solutions that satisfy the problem constraints. Then, the optimization problem consists in finding the element $\mathbf{x}=\mathbf{x}^* \in \mathbf{X}$ that minimizes the $f_c(\mathbf{x})$ function. Mathematically, it can be expressed as follows:

$$\mathbf{x}^* \in \mathbf{X} \quad \text{and} \quad f_c(\mathbf{x}^*) \leq f_c(\mathbf{x}), \forall\, \mathbf{x} \in \mathbf{X} \tag{2}$$

In order to employ the FP algorithm to tuning the AC parameters, the following points must be defined: the cost function, the variables being optimized and the constraints defining "admissible solutions", i.e., the admissible limits for the variables being optimized.

In order to determine the cost function it is necessary to define the per-flow VoIP and the AF traffic requirements. Because this work considers a single node AC, these requirements are expressed in terms of the delay and the packet loss caused by the node. The per-flow VoIP requirement is expressed as the percentile of packets that can exceed a delay bound or be dropped in a single flow. For example, a QoS requirement (97th , 50ms) defines that a maximum delay of 50ms must be observed by 97% of the packets of each individual flow, i.e., only 3% of the packets can violate the QoS requirements by exceeding the 50ms limit or being dropped. The AF traffic requirement is imposed in terms of the overall packet-loss ratio. We have limited the AF queue to a size corresponding to the maximum admissible delay assigned to the AF packets (i.e, the single node delay contribution), so all packets that exceed this delay bound are dropped.

The main idea of the cost function is to induce the optimization process to find a solution for the admission controller that maximizes the number of flows served without violating the QoS requirements of both, VoIP and the AF data traffic. Considering an evaluation period, let F_a, $\overline{F}_a$, $\widehat{F}_a$ and F_v be, respectively, the absolute number of accepted flows, the average number of active flows, the estimated maximum number of active flows and the absolute number of admitted flows that violate the QoS requirements. We also define AF_d as the percentage AF packets dropped with respect to the total number of packets marked as AF. The cost function can, then, be represented as:

$$f_c(\mathbf{x}) = \left[1 - \frac{\overline{F}_a}{\widehat{F}_a}\right] + \left[\varphi \frac{F_v}{F_a}\right] + \left[\phi\, AF_d\right] \tag{3}$$

The **x** vector represents the AC variables that must be optimized. Because we have assumed a simple PBAC controller, in this case, the variable is the maximum number of simultaneously active flows. The $\widehat{F}_a$ term is not required to be precise, as it works only as a normalization factor. It can be computed using the techniques suggested by [15]. The second component introduces a penalization for violating the QoS requirements. The φ [undimensional] factor defines the "penalty" for admitting a violated flow. The third term defines a penalization for dropping the AF traffic. The penalization weight is controlled by the ϕ [undimensional] factor. Both φ and ϕ are inputs to the optimization process and depend on the service provider business policy.

Extremely high occupation rates that induce too many VoIP or AF violations will be penalized by the cost function. Therefore, no special treatment is required for avoiding unfeasible solutions.

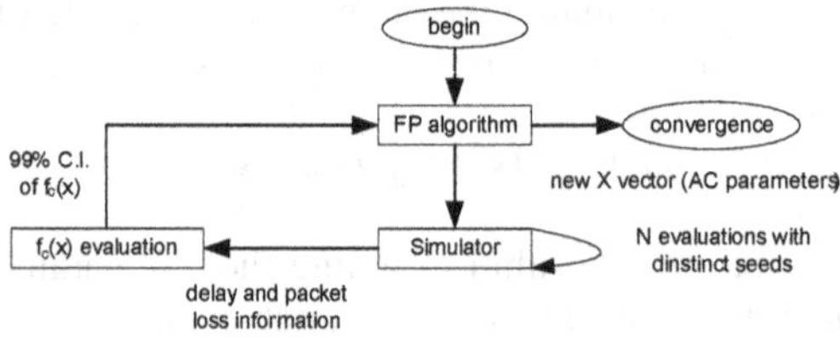

Fig. 3. Optimization Flow

Figure 3 illustrates the complete optimization flow, where a simulator and the FP method are combined. The simulator is activated for computing the cost function of each new vertex of the polyhedron. In order to reduce the seed influence, N simulations with distinct seeds are computed for each new vertex generated by the FP method. During a simulation, the AC parameters are tested against a variable load scenario that follows a "peaked call arrival pattern" shared with the self-similar data traffic, as described in section 3. In this paper we have employed N=10. The cost function (f_c) is computed for each seed, and a 99% confidence interval for the f_c average is determined by assuming a t-distribution. The cost assigned to a PBAC solution corresponds to the worst bound in this interval (i.e., the highest cost function).

The AC parameters obtained using the optimization approach are valid for a specific scenario defined by the link capacity, the per-flow delay/packet-loss bound, the leaky bucket parameters, the per-flow VoIP percentile performance and the "φ" and "ϕ" parameters.

Initially, the simulations were performed using the NS-2 [16]. However, the NS-2 was not scalable enough for supporting the most complex simulation scenarios (usually, hundreds of simulations are required before the FP convergence). Therefore, a simulator oriented to discrete events based on the actor/message paradigm [17] was implemented using the C language. In this case, the network elements, the VoIP entities and the self-similar source were implemented as actors by using threads, and the packets were implemented as messages exchanged between the actors. For the simpler scenarios, the results of the C simulator have been compared against the NS-2, leading to identical results.

5 Simulation and Results

In this section, we have evaluated our approach by optimizing the PBAC method for distinct link capacities and distinct leaky bucket parameters. In all scenarios, the VoIP offered load followed the template defined in Figure 1, but the conditions of the data traffic were modified by adjusting the leaky bucket parameters: service rate and bucket size. Figures 4 to 6 have considered a scenario that imposed a per-flow delay bound of 50ms for the 97th percentile of the VoIP packets and a delay bound of 250ms for the aggregate of AF packets. The performance parameters were adjusted respectively to $\phi = \varphi = 10$. This represents a strong penalization for violating VoIP or AF guarantees. As a consequence, in all scenarios, there was no violation of VoIP flows or AF traffic. Therefore, this information was omitted in the figures.

Both, the self-similar and the VoIP traffic can significantly vary according to the seed values. To avoid misinterpretation, the results presented in this section correspond to the average of 30 simulations with distinct seeds, presented with a 99% confidence interval. Evaluating the approach with so many distinct seeds assures that the simulated results achieved by the controller will be robust enough to be reproduced in the real world.

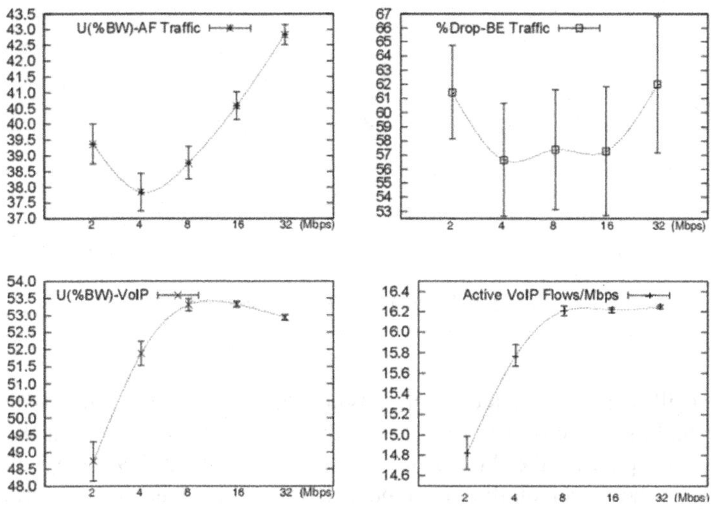

Fig. 4. Effect of the Link Capacity

The left side of Figure 4 illustrates the effect of the link capacity on the share of bandwidth occupied by the AF and the VoIP traffic. The right side of the figure illustrates the effect of the link capacity on the BE traffic drop rate and the average number of active flows per Mbps. One observes in the figure that the number of VoIP flows per Mbps increases with the link capacity. As a result, the share of the link bandwidth occupied by the VoIP traffic also increases with the link capacity. This effect is justified because a longer VoIP queue can be tolerated in terms of packets when a fixed maximum delay is imposed to a node with higher service rate. This effect is also observed for the AF traffic. Also, note the smoothing effect obtained by aggregating more VoIP flows, as observed by the reduced dispersion around the active VoIP flows average.

Figure 5 illustrates the effect of modifying the leaky bucket rate (expressed as a % of the link capacity). In this scenario, the link capacity was fixed in 8 Mbps. Note that increasing the leaky bucket rate increases the offered load of AF packets. One observes that the higher priority VoIP traffic is limited by the PBAC controller in order to accommodate the additional AF traffic. In all evaluations, the average bandwidth of the self-similar traffic (i.e., offered load before marking) was fixed in 50% of the link capacity (*m* parameter). However, increasing the bucket rate above 50% still affects the traffic, due the fBm variation.

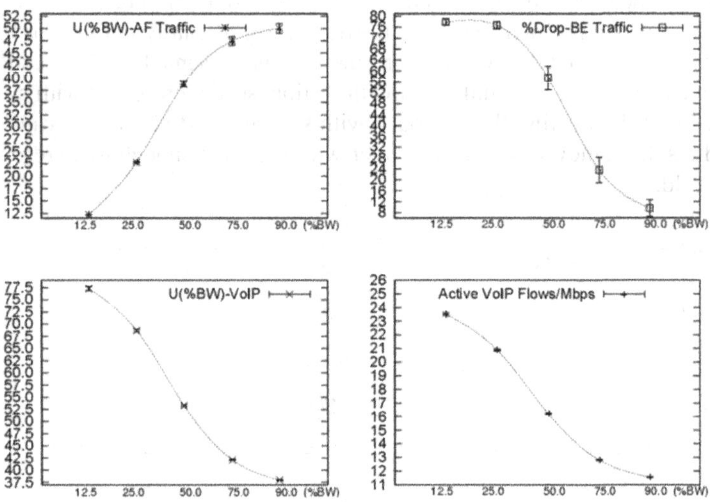

Fig. 5. Effect of Leaky Bucket Rate

Figure 6 illustrates the effect of modifying the leaky bucket size. Again, in this scenario, the link capacity was fixed in 8 Mbps. Note, in Figure 4, that the share of bandwidth occupied by the AF traffic was always less than 50% (the leaky bucket rate). This is the effect of the small bucket size adopted in this scenario (only 2 MTU). Increasing the bucket size enable to accommodate the self-similar traffic variation, increasing the AF traffic rate. Consequently, the AF traffic shaped by the

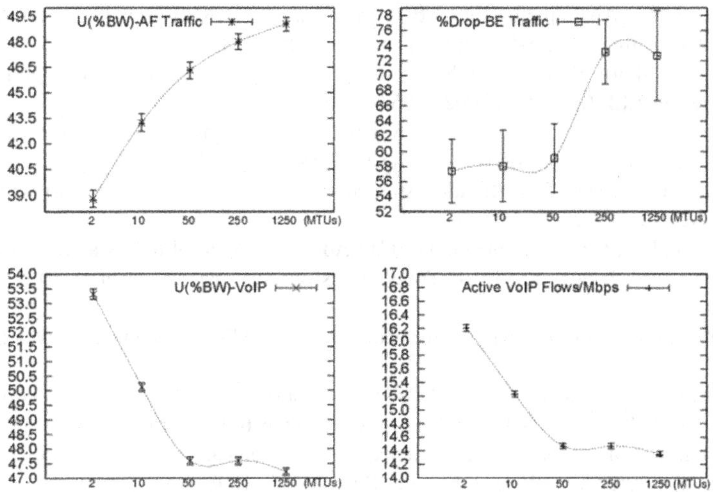

Fig. 6. Effect of the Leaky Bucket Size

leaky bucket is not as well-behaved as the traffic in Fig. 4. In this case, the PBAC had to reduce the number of VoIP flows per Mbps accordingly.

6 Conclusion

This paper has presented an optimization method that helps the design of DiffServ AC algorithms capable of providing per-flow delay/packet-loss guarantees. The approach was illustrated considering a scenario where a node link was shared by distinct types of traffic: VoIP and self-similar traffic marked by a leaky bucket algorithm. A typical DiffServ node capable of scheduling only large classes of traffic (i.e., flow-unaware) was studied. Because most results obtained in the literature target only the performance of aggregated traffic, there is no accurate analytical model capable of treating a per-flow scenario with this type of scheduler.

The method described in this paper was capable of achieving the required performance results in all scenarios, including distinct link capacities and variations on the level of self-similar packets marked as assured forwarding. The drawback of this method, however, is that the results obtained to a PBAC are valid only to a specific link capacity and specific leaky bucket parameters. The controller, however, is robust with respect to the offered load variation.

Future publications will explore the method described in this paper for optimizing MBAC algorithms and designing AC controllers for multi-hop scenarios.

References

1. Blake, S., Black, D., Carlson, M., Davies, E., Wang, Z., Weiss, W.: Architecture for differentiated service. IETF RFC 2475 (December 1998)
2. Breslau, L., Jamin, S., Shenker, S.: Comments on the performance of measurement-based admission control algorithms. INFOCOM 2000 3, 1233–1242 (2000)

3. Moore, A.W.: An implementation-based comparison of Measurement-Based Admission control algorithms. Journal of High Speed Networks 13, 87–102 (2004)
4. Siripongwutikorn, P., Banerjee, S.: Per-flow delay performance in traffic aggregates. Proceedings of IEEE Globecom 2002 3, 2634–2638 (2002)
5. Xu, Y., Guerin, R.: Individual qos versus aggregate qos: A loss performance study. IEEE/ACM Transactions on Networking 13, 370–383 (2005)
6. Knightly, E.W., Shroff, N.B.: Admission control for Statistical QoS: Theory and Practice. IEEE Network 13, 20–29 (1999)
7. Spitler, S., Lee, D.C.: Optimization of Call Admission Control for a statistical multiplexer allocating link bandwidth. IEEE Transactions on Automatic Control 48(10), 1830–1836 (2003)
8. Cisco Systems. Traffic Analysis for Voice over IP. Cisco Document Server (Setember 2002)
9. Norros, I.: On the use of fractional Brownian motion in the theory of connectionless networks. IEEE Journal on Selected Areas in Communications 13(6), 953–962 (1995)
10. Hosking, J.R.M.: Modeling persistence in hydrological time series using fractional brownian differencing. Water Resources Research 20, 1898–1908 (1984)
11. Boutremans, C., Iannaccone, G., Diot, C.: Impact of link failures on VoIP performance. In: Proceedings of 12th NOSSDAV workshop, pp. 63–71. ACM Press, New York (May 2002)
12. Jiang, Y., Yao, Q.: Impact of FIFO aggregation on delay performance of a differentiated services network. In: Kahng, H.-K. (ed.) ICOIN 2003. LNCS, vol. 2662, pp. 948–957. Springer, Heidelberg (2003)
13. Nelder, J.A., Mead, R., Simplex, A.: Method for Function Minimization. Comput. J. 7, 308–313 (1965)
14. Lagarias, J.C., Reeds, J.A., Wright, M.H., Wright, P.E.: Convergence Properties of the Nelder-Mead simplex method in low dimensions. SIAM. J. Optim. 9(1), 112–147 (1998)
15. Guérin, R., Ahmadi, H., Naghshineh, M.: Equivalent capacity and its application to bandwidth allocation in high-speed networks. IEEE Journal on Selected Areas in Communications 9(7), 968–981 (1991)
16. NS-2. The ns-2 simulator. Available at: http://www.isi.edu/nsnam/ns/
17. Agha, G.: Actors: A Model of Concurrent Computation in Distributed Systems. MIT Press, Cambridge. Available at: http://hdl.handle.net/1721.1/6952

Availability-Aware Multiple Working-Paths Capacity Provisioning in GMPLS Networks

Han Ma[1], Dalia Fayek[1], and Pin-Han Ho[2]

[1] School of Engineering, University of Guelph, Guelph, Ontario, Canada
[2] Electrical and Computer Engineering, University of Waterloo, Waterloo, Ontario, Canada

Abstract. Network protection mechanisms are critical in the design of IP-based networks. In GMPLS networks, backup path protection and Shared-backup protection have been widely studied. More recent protection mechanisms, Self-Protecting Multipath (SPM), can be implemented in GMPLS networks. In this paper, we evaluate the traffic provision in SPM GMPLS networks where events of up to two simultaneous link failure can occur. We present a mathematical formulation to perform optimal capacity allocation in an SPM network environment. Network Service Provider (NSP) could use this mathematical model to design a network with certain availability requirement according to Service Level Agreement(SLA) of each connection requirement in dual link failure scenarios.

1 Introduction

With the growth of real-time multimedia applications and mission-critical applications, service availability becomes a key issue in IP networks, especially in the backbone networks. Service Level Agreement (SLA) defined a standard service level for Network Service Providers (NSP) with possible revenue reward and penalty. Network restoration and protection mechanisms are widely used to provide higher network or service availability. Network protection mechanisms achieve a better result than restoration [1], [2]. Protection paradigms vary from traditional dedicated backup path protection [3] to shared backup path protection [4]. Most protection mechanisms include a working path and one backup path which could be dedicated or shared with others [5]. One dedicated backup path can provide 100% availability for one link failure, but cannot reach full availability for two or more simultaneous link failures. The engineering of working and backup paths normally results in more backup bandwidth but also in more cost.

Network Service Providers face two major control issues: (1) provide higher network availability to guarantee performance for real-time applications and good end-user experience and (2) cut on cost. Since low cost is the key to success in business, finding a low cost way to provide high network protection level is sought by NSPs.

D. Medhi et al. (Eds.): IPOM 2007, LNCS 4786, pp. 85–94, 2007.

Network protection mechanisms are available at many network layers: physical link, optical transport network, SDH/SONET networks and IP/MPLS networks [6,7,8]. Today's network backbone is widely based on IP/MPLS architecture. Network protection in IP/MPLS layer can provide high level flexibility and better resource efficiency. Research presented in [9,10] are examples about the protection trends in IP/MPLS networks where it was shown that Shared Backup Protection Paths (SBPP) can provide flexible, low cost and end-to-end protection.

The rest of this paper is organized as follows. Section 2 describes recent mechanisms of protection based on Self-Protecting Multipath (SPM). Section 3 introduces a linear programming model based on double link failure with availability requirements. This LP model is of interest to NSP's since it provides a solution for traffic provision in SPM network under double link failure situation. It also provides a way to estimate link capacity requirement and design backbone network for NSP to meet SLA requirement with low cost. In Section 4, we present and discuss simulation results for the proposed LP model. Concluding remarks are presented in Section 5.

2 Self-Protecting Multipath Background

Today's backbone topology tends to increase in complexity. In most cases, there are many (more than two) paths to connect two network nodes. In GMPLS provision networks, it is possible to setup two or more working paths. If working paths have spare capacity, they could work as backup paths to other paths when network failure occurs. Self-Protecting Multipath (SPM) is a recent network protection mechanism [11] that is designed for GMPLS multiple working path establishment. In SPM networks, network traffic is split through multiple disjoint paths in normal working status [12], [13]. Fig. 1 depicts an example SPM layout where between nodes A and B, there are three disjoint working paths: A1, A2 and A3. The three working paths together provide the total bandwidth requirement for sessions between A and B. If there is (are) failure(s) in one (or two) of these multiple working paths, the traffic is re-allocated to working path(s) which works under current network failure (if there is enough spare network capacity along the working path(s)). This mechanism provides a certain level of network protection when there is enough spare capacity.

In this paper, we focus on capacity provision in a multipath GMPLS network to meet certain service level agreement requirement when the network has possibly up to two link failures happening simultaneously. In this research, each additional link incurs a certain cost. Our aim is to find a minimum cost solution based on optimized capacity provisioning both in normal working status and failure situations. The contribution of this paper is in devising a mathematical model that takes into account more than one failure scenario. This mathematical model could help NSP to design or plan their network (determine the minimum required network link cost) to reach SLA in double network failure situation according to certain traffic requirement or potential traffic requirement. As we

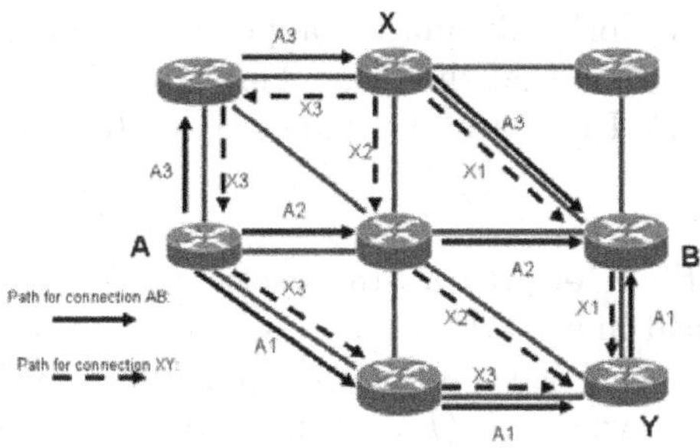

Fig. 1. Example of a Self-Protecting Multipath network

will show, more than two simultaneous link failures result in a non-tractable complexity for the proposed algorithm to find a solution in polynomial time.

3 Mathematical Model for Multiple Disjoint Path Protection

A network with multiple paths has two or more physically disjoint working paths for each source-destination pair. That is, disjoint paths are paths that do not share intermediate nodes nor physical links. Each working path is formed by one or more physical link, and each physical link can support one or more working path. A link's capacity poses an upper limit on the maximum bandwidth of all the working paths using this link. It also determines the spare capacity which can be used for backup paths. In network design phase, routing strategy and bandwidth usage in every link are determined by typical traffic loads and their (re)distribution during normal (failure) operating conditions. We present a linear programming model to optimize bandwidth allocation among multiple working and shared backup paths in 3.1 and 3.2. Then we analyze the complexity of the linear program model in 3.3.

3.1 Problem Formulation

First of all, we formulate this problem under the following assumptions.

- We assume that protection is only provided in the IP and GMPLS layers (i.e., network and transport layers).
- We consider link failures as the only source of failure in this research. Although both physical links and network devices (such as routers in every network node) could contribute to the network failure, however redundancy design in network devices is common, that is why it is excluded from our model.
- We do not consider the effect of shared risk link group (SRLG). SRLG is a set of links in one network that could fail simultaneously. In most cases, group failure is due to either these links are all in one physical location or these links differ only in wavelength but are physically carried in one fiber.

- We assume that network link failures happen independently and constitute a memoryless process and that the Mean Time To Failure (MTTF) and Mean Time To Repair (MTTR) are also independent (exponentially distributed random variables).

Network and Traffic Definition. In a network $G = (V, E)$, there are $|V|$ nodes and $|E|$ physical links (Edges). We use $c \in C$ to denote connection requirements incident on G. Each connection requirement c is described by a tuple $< S_c, D_c, A_{c,SLA}, T_c >$, where S_c (D_c) is the source (destination) node of connection c; $A_{c,SLA}$ is the availability requirement of connection c according to the service level agreement (SLA) (in percentage bandwidth); T_c is the bandwidth requirement of connection c (in Mbps).

Path-Link Matrix. Each connection $c \in C$ has one set of paths B_c. We denote each disjoint path in B_c as b_c. For each connection $c \in C$, the matrix P_c describes the membership of physical links in disjoint paths. For network G that has $|E|$ physical links, matrix $P_c = \{0,1\}^{|B_c| \times |E|}$. Therefore, for $\forall b_c \in B_c$ and $j = 0, \ldots, |E| - 1$, set $p_{b_c,j}$ as an entry in matrix P_c:

$$p_{b_c,j} = \begin{cases} 0 \text{ if path } b_c \text{ does not use link } j \\ 1 \text{ if path } b_c \text{ uses link } j \end{cases}$$

Failure pattern. In this work, we consider up to two simultaneous link failures (Sec. 3.1). Let F denote the set of all possible failure patterns. and f_i denote one such failure pattern, with f_0 being the no-failure situation. Let also $f_{i,j}$ denote whether there is a failure in physical link j under failure pattern i. Then for $i = 0, \ldots, |F| - 1$, $j = 0, \ldots, |E| - 1$:

$$f_{i,j} = \begin{cases} 0 \text{ if there is no failure in link } j \text{ in pattern } i \\ 1 \text{ if there is a failure in link } j \text{ in pattern } i \end{cases}$$

Since we only consider up to two simultaneous link failures in network G, $f_i (i \neq 0)$ can at most have two entries equal to 1, and the stationary distribution of f_i is given by $\pi(f_i)$.

3.2 Linear Programming Model

Free variable. In multiple path networks, for each connection $c \in C$, there are $|B_c|$ different self-protecting working paths to provide a total of T_c bandwidth requirement that is dynamically allocated within the B_c paths. In case of failure(s), the bandwidth provision has to be reconfigured to guarantee the total bandwidth availability at the $A_{c,SLA}$ level requirement.

We use the vector variable $\theta_{b_c}(f_i) \in [0,1]^{B_c}$ as the protection level for each disjoint path b_c for connection $c \in C$ under failure pattern $f_i \in F$. The protection level is described as percentage of bandwidth distribution among the disjoint paths of connection c.

Constraints. First we define the constraint on the value of the free variable $\theta_{b_c}(f_0)$ when there is no failure in any link, denoted by the pattern f_0 . The total traffic requirement T_c of connection c must be 100% provisioned in disjoint working paths:

$$\sum_{b_c \in B_c} \theta_{b_c}(f_0) = 1 \quad \forall c \in C \tag{1}$$

In case of failure f_i, $i \neq 0$

$$\sum_{b_c \in B_c} \theta_{b_c}(f_i) \leq 1 \qquad \forall c \in C, \forall f_i \in F \tag{2}$$

Equation 2 states that the total bandwidth distributed among working paths $b_c \in B_c$ cannot exceed the traffic requirement T_c for all connections $c \in C$ but might be less due to failure.

Since no traffic is transported in a link with failure, we use the P_c matrix to transfer link failure to path failure; hence the path failure is captured by the third constraint (Eq. 3) whose corresponding free variable θ_{b_c} must be zero:

$$\theta_{b_c}(f_i) \sum_{j=0}^{|E|-1} p_{b_c,j} \cdot f_{i,j} = 0 \quad \forall c \in C, \forall f_i \in F \tag{3}$$

When failure occurs, the protection level for connection c under failure pattern f_i is given by $\sum_{b_c \in B_c} \theta_{b_c}(f_i)$. Now, constraint 4 states that the average protection level should exceed the availability requirement of connection c:

$$\sum_{i=0}^{|F|-1} \pi(f_i) \cdot \left(\sum_{b_c \in B_c} \theta_{b_c}(f_i) \right) \geq A_{c,SLA} \quad \forall c \in C \tag{4}$$

Finally, the fifth constraint ensures that every free variable $\theta_{b_c}(f_i)$ has a value between 0 and 1:

$$0 \leq \theta_{b_c}(f_i) \leq 1 \quad \forall b_c \in B_c, \forall c \in C, \forall f_i \in F \tag{5}$$

Objective. The objective of this LP problem is to minimize the total cost in network G. This cost is defined as follows. Let w_j be the cost in dollars/Mbps for link j, and define $S_{i,j}$ to be the cost of link j (in dollars) under failure pattern i and traffic distribution $\theta_{b_c}(f_i)$ given by:

$$S_{i,j} = w_j \sum_{c \in C} \sum_{b_c \in B_c} p_{b_c,j} \cdot \theta_{b_c}(f_i) \cdot T_c$$

The cost of link j can be expressed as:

$$C_j = \max_{i=0}^{|F|-1} S_{i,j} \quad \forall j \in E$$

The objective function of this LP model can now be expressed as:

$$\min \sum_{j=0}^{|E|-1} C_j \tag{6}$$

3.3 Complexity Analysis

The number of free variables and constraints determine the complexity of an LP model. The number of all possible single and double link failure patterns are $|E|$ and $|E|(|E|-1)/2$, respectively. Therefore, the total number of failure patterns is equal to $|F| = 1+|E|(|E|+1)/2$. For a total number of $|C|$ traffic connections, the number of free variables $\theta_{b_c}(f_i)$ is: $[1+|E|(|E|+1)/2] \times \max_{c\in C} |B_c| \times |C|$.

For the five constraints in this LP model, the total number of equations for constraint (C1) and (C4) is $|C|$ equations, each. (C2) and (C3) have $|C| \times [1 + |E|(|E|+1)/2]$ equations each. C5 has a total of $[1+|E|(|E|+1)/2] \times \max_{c\in C} |B_c| \times |C|$ equations.

From this analysis, we can see that the complexity of this problem is related to the number of links, connection requirement and maximum number of disjoint paths, the number of links being more dominant in the complexity order, that is, $O(|E|^2)$. For more than 2 simultaneous link failure scenarios (e.g. n), the complexity becomes $O(|E|^n)$. The likelihood of n simultaneous failures significantly decreases with n, and mathematical complexity increases in $O(|E|^n)$. Therefore, for practical purposes, this study limits the number of simultaneous failures to two for mathematical tractability. This makes this LP model practically manageable for NSP's to configure bandwidth allocations in backbone networks and in whole sale businesses. Simulations, discussed next, show that this model is computationally tractable so far when the number of edges reaches a total of 50.

4 Simulation and Results

The simulation is aimed at studying the efficiency of the availability-aware multipath capacity provision model in double network failure situations with a variety of protection levels. This numerical study also demonstrates the computation time of the proposed optimization model.

4.1 Efficiency of Availability Model

We investigate the proposed LP model in selective networks with specific topological characteristics: all networks in this simulation contain at least 3 disjoint paths for each connection requirement (not for each SD pair).

In this study, and without loss of generality, we assume that $w_j = 1$ for links. We also assigned same availability requirement $A_{c,SLA}$ to each connection $c \in C$. We assume that the link failure rate is proportional to the length of the link. The average failure rate 1/MTTF per kilometer is set to 0.00005. Further we set all links to be 100 kilometers long.

Fig. 2 shows one network with 30 edges and 13 nodes which is used in this simulation to validate our mathematical model developed in Section 3. In this simulation, the traffic incident on the network shown consists of 20 connection requirements between random node pairs. Each connection has at least 3 disjoint paths.

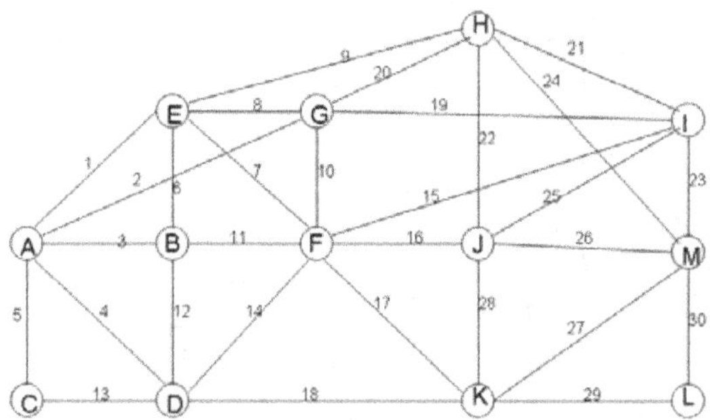

Fig. 2. One sample network with 30 edges used in simulation

Fig. 3 shows total network capacity and spare capacity vs. connection availability. We notice it requires more spare capacity for higher protection levels. In SPM networks, link capacities are dynamically distributed in every working paths. Trying to calculate the spare capacity, we first calculate the network capacity for shortest path routing without failure, then we calculate total required capacity with double failure protection. We consider the difference between these two as spare capacity.

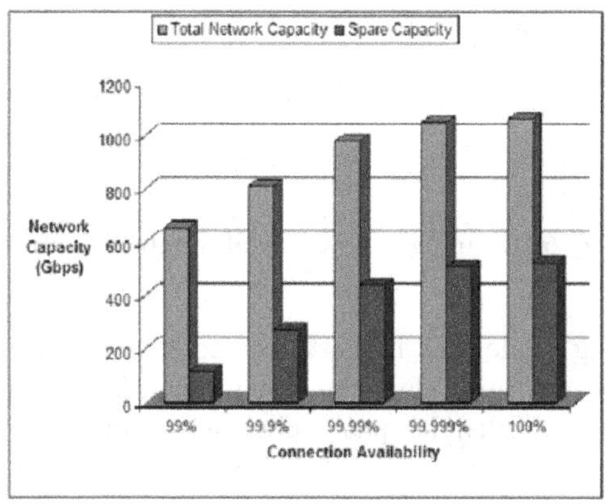

Fig. 3. Spare capacity vs. Connection Availability

The spare capacity percentage is not only affected by connection availability, but also by the number of disjoint paths. In Fig. 4, simulation results for a 30 edges network and 20 connections is shown where it can be seen that high connection availability leads to higher percentage of spare capacity. When we limit the use of disjoint paths for each connection, it requires even more spare capacity. When we only allow 2 disjoint paths, the network cannot reach high connection availability. This result demonstrates that the proposed availability-aware

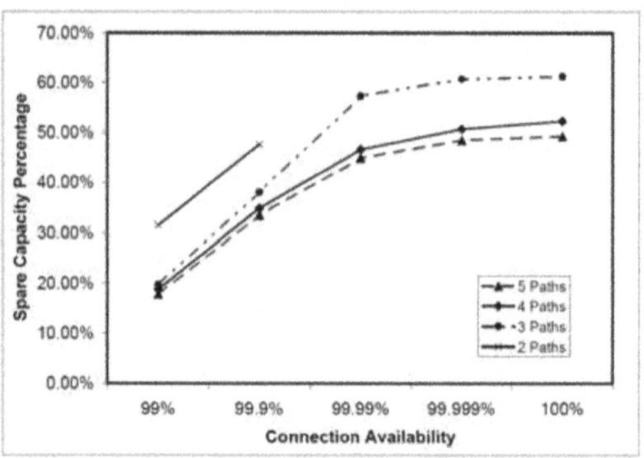

Fig. 4. Spare capacity percentage affected by connection availability and number of disjoint path

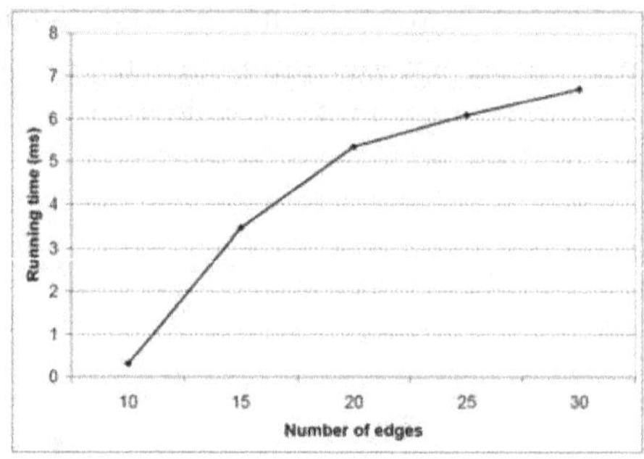

Fig. 5. Running time vs. total number of edges

capacity provisioning model has higher efficiency when more disjoint paths are used for each connection and that there is no significant gain in performance when moving from 4 to 5 disjoint paths per connection.

4.2 Performance Analysis

Here we present the performance of the proposed LP model in terms of computation time. As discussed in Section 3.3, the complexity is $O(|E|^2)$. To illustrate the relationship between network edge and running time, we logged the run-time for different number of edges, from 10 to 30 edges (Fig. 5). These simulations used ILOG Cplex version 10.1 as the LP solver; and Microsoft C#.Net as development environment. All simulations are run on a Dell P650, 2.8GHz CPU and 1GB RAM under Microsoft Window XP Pro. The test results show the running time increases quadratically with $|E|$.

5 Conclusion

In this paper, we presented a Linear Programming optimization model to configure working and backup capacity allocation for multiple disjoint paths in self-protected networks. This model minimizes the total cost for some given connection profiles with SLA requirements in up to two simultaneous link failure situations. The linear constraints and number of free variables in this model where shown to constitute an $O(|E|^2)$ complexity for the LP solver.

The current network simulations are now conducted with higher number of edges and with noticeable success with our model. There are some fine tunings that need to be addressed. For future work, we will explore the following: (*i*) precise modeling of the failure patterns' stationary distribution, (*ii*) present more analysis for other network topologies, (*iii*) compare this multiple working path model with shared backup model, and (*iv*) add physical network link capacity constraint to current model, so that the applicability of this model can span not only the network planning stage but also to upgrade and/or allocate traffic in exiting network.

As presented in this paper, this model is adequate to provide guidelines for NSP's in planning their core network link capacity with working and spare capacity allocation under the statistically defined failure patterns with fix network topology.

References

1. Zhang, J., Zhu, K., Zhan, H., Mukherjee, B.: A new provisioning framework to provide availability guaranteed service in WDM mesh networks. In: IEEE International Conference on Communications, pp. 1484–1488 (May 2003)
2. Autenrieth, A., Kirstadter, A.: Engineering End to End IP Resilience Using Resilience-Differentiated QoS. IEEE Communication Magazine, 50–57 (January 2002)
3. De Patre, S., Maier, G., et al.: Design of Static WDM Mesh Networks with Dedicated Path-Protection. In: IEEE International conference on Communications (June 2001)
4. Ho, P.-H., Tapolcai, J., Mouftah, H.T., Yeh, C.-H.: Linear Formulation for Path Shared Protection. In: IEEE International Conference on Communications, pp. 1622–1627 (June 2004)
5. Choi, H., Subramaniam, S., Choi, H.-A.: On Double-Link Failure Recovery in WDM Optical networks. In: WDM Optical Networks, IEEE INFOCOM, IEEE Computer Society Press, New York (June 2002)
6. Liu, Y., Tipper, D., Siripongwutikorn, P.: Approximating optimal spare capacity allocation by successive survivable routing. IEEE/ACM Trans on Networking 13(1), 198–211 (2005)
7. Ho, P.-H., Tapolcai, J., Cinkler, T.: Segment Shared Protection in Mesh Communications Networks with Bandwidth Guaranteed Tunnels. IEEE/ACM Transactions on Networking 12(6), 1105–1118 (2004)
8. Cao, J., Guo, L., Yu, H., Li, L.: Partial SRLG-disjoint shared path protection with differentiated reliability in survivable WDM network. International Journal of Electronics and Communications (AEU) 61, 353–362 (2007)

9. Amin, M., Ho, K.-H., Pavlou, G., Howarth, M.: Improving Surviability through Traffic Engineering in MPLS Networks. In: Proceedings of the 10th IEEE Symposium on Computers and Communications, 27-30, pp. 758–763 (2005)
10. Ho, P.-H., Mouftah, H.T.: Reconfiguration of Spare Capacity for MPLS-Based Recovery in the Internet Backbone Networks. IEEE/ACM Transactions on Networking 12(1), 73–84 (2004)
11. Menth, M., Reifert, A., Milbrandt, J.: Self-Protecting multipaths - A simple and resource efficient protection switching mechanism for MPLS network. In: 3rd IFIP-T6 Networking Conference, pp. 526–537 (May 2004)
12. Menth, M., Martin, R., Spoerlein, U.: Network Dimensioning for the self-protecting multipath: A performance study. In: IEEE International Conference on Communications, pp. 847–851 (June 2006)
13. Menth, M., Reifert, A., Milbrandt, J.: Comparison of Capacity Requirements for the Self-Protecting Multipath and Similar Mechanis in Resilient Packet Networks. IEEE Next Generation Internet Design and Engineering, 300–309 (2006)

Distributed Dynamic Protection of Services on Ad Hoc and Peer to Peer Networks

Jimmy McGibney and Dmitri Botvich

Telecommunications Software & Systems Group,
Waterford Institute of Technology, Waterford, Ireland
{jmcgibney,dbotvich}@tssg.org

Abstract. A collaborative system for dynamic refinement of security in peer-to-peer and mobile ad hoc networks is described in this paper. This is based on a closed loop system where live distributed trust measures are used to modify access control settings in a changing threat environment. A service oriented trust overlay architecture and model underlies this system. In this model, services have associated trust thresholds – the more sensitive the service, the higher the threshold. The results of simulations of the dynamics of this kind of system are presented and a variety of algorithmic approaches to managing trust are analysed and discussed. It is demonstrated that this dynamic system has the potential to enhance security and access control efficiency and that it displays properties of robustness when faced with malicious entities that attempt to corrupt the system.

Keywords: Intrusion detection, distributed trust management, network security.

1 Introduction

With the advent of mobile ad hoc and other unstructured peer-to-peer systems, existing approaches to security are no longer sufficient. It cannot be assumed, for example, that participants always have access to centralised resources such as a public key infrastructure to validate credentials. At the same time, traditional perimeter security is of little benefit. The risks are substantial with the use of shared radio spectrum, and there is a history of weak wireless security protocols.

In this paper, we present an approach to dynamic management of security that is based on distributed trust. Nodes offer services and peer nodes use them. Access to some such services require more trust than others. Having certain services enabled on a node could expose that node to various types of attack, even though it might have good protection mechanisms in place and be kept up to date with patches as new exploits are common and it is easy for users to misconfigure systems and services.

How trust information is shared and how it is handled by individual nodes is of course critical. We consider algorithms for sharing trust information according to individual nodes' preferences and also for nodes to effectively update trust. This needs to take into account to risk of some "bad" nodes deliberately attempting to corrupt the system, possibly in collusion with each other.

D. Medhi et al. (Eds.): IPOM 2007, LNCS 4786, pp. 95–106, 2007.

For the purposes of this work, we adopt a two layer model for communications between nodes. Nodes can either interact for service usage or to exchange trust information. For modelling purposes, each service usage interaction is a discrete event. As we will see in section 3, a logically separate trust management layer handles threat notifications and other pertinent information.

The remainder of this paper is organised as follows. The next section describes related work. Section 3 then outlines our trust overlay architecture and this is followed in section 4 with a detailed model of our system. Section 5 presents and discusses results of initial simulations to evaluate the system. Section 6 concludes the paper and discusses possibilities for future work.

2 Related Work

A significant body of literature has emerged in the recent years on trust and reputation systems. Much of this work attempts to model trust computationally so that it can be used in making decisions related to electronic transactions, in a sense mimicking people's well-evolved forms of social interaction. A recent paper by Jøsang et al. [1] surveys the state of the art.

Many open issues exist and challenges remain. The social concept of trust is very complex and sophisticated, perhaps deceptively so. Trust is closely related to many other social concepts such as belief, confidence, dependence, risk, motivation, intention, competence and reliability [2]. It is also interwoven with the areas of accountability [3] and identity [4].

There are currently several different definitions and interpretations of what trust means and how it can be used. In some cases, trust is modelled as a probabilistic measure indicating confidence in a certain type of behaviour [5], and this measure is used as a basis for deciding whether to rely on another entity. In other cases, "to trust" is taken to mean making the decision itself. In our work, we model trust as a vector of measures to allow for a variety of trust-related factors that might influence a decision, such as a score based on experience and reputation, service-specific scores, confidence in these scores and a measure of their recency (as trustworthiness might be expected to decay over time).

Several authors have proposed reputation systems specifically for peer-to-peer networks. Most of these systems either rely on a global view of reputation, via a centralised server, or attempt to estimate the global view by polling a certain number of other peers. For example, the *secure EigenTrust* algorithm [6] polls a set of M peers that are designated as score managers for each peer i. Our system differs from this in that we do not have any such designation of peers, and thus we avoid having to maintain complex management information. Likewise, Gupta et al. [7] propose a partially distributed system that depends on reputation computation agents. Damiani et al. [8] propose a more truly distributed system, though it depends on polling peers and computing reputation each time a peer tries to download a resource. Other systems, like Google's *PageRank* [9] are essentially centralised and peers do not directly participate at all. Repantis & Kalogeraki have recently proposed a decentralised reputation system [10].

Our system differs from these in that all collaborating peers can make referrals equally, and the trusting node makes a subjective determination of how to weight these – usually based on its level of trust in each recommending node, its recency, and a requirement for corroboration. In fact, different nodes may have their own various strategies for updating trust and making referrals. Crucially, there is no dependence on specific nodes (such as score managers) being always switched on or within range.

Several authors have also looked more specifically at trust in wireless ad hoc networks. The usefulness of distributing trust in this situation was identified by Zhou & Haas [11] and more generally for ubiquitous services by Stajano & Anderson [12]. Much of the focus to date on securing ad hoc networks has been on the protection of routing, and often on specific protocols. Yang et al. [13], for example, propose a collaborative solution that is applicable to a variety of routing and packet forwarding protocols. This system is reactive, dealing dynamically with malicious nodes. Reputation systems for mobile ad hoc networks have also been proposed (e.g. by Buchegger & Le Boudec [14]).

3 Trust Overlay

3.1 Representing Trust Information

In an ad hoc or truly peer to peer network, there is no way to store trust information centrally. In any case, trust by its nature is subjective and thus best managed by the entity doing the trusting. This decentralised trust management is the essence of distributed trust.

Rather than having binary off/on trust, as exists in typical public key infrastructures, we would prefer to more accurately mimic social trust by setting a more fuzzy trust level. Different services can then make appropriate decisions based on this trust level – e.g. certain actions may be allowed and others not. In our system, we model trust as a vector. In the simplest case, at least if there is just one service, this can be viewed as a simple number in the range (0,1). Each node may then maintain a local trust score relating to each other node of which it is aware. If node A's trust in node B is 1, then node A completely trusts node B. A score of 0 indicates no trust. Note that this is consistent with Gambetta's definition of trust as "a particular level of the subjective probability with which an agent assesses that another agent or group of agents will perform a particular action, both before he can monitor such action and in a context in which it affects his own action" [5]. If trust is to be a probability measure, then the (0,1) range is natural.

3.2 Architectural Overview

As in [15], we propose the overlay of a distributed trust management infrastructure on top of the service delivery infrastructure. With this architecture, a trust management overlay layer operates separately from all aspects of service delivery and usage. Two message passing interfaces are defined between the service usage layer and the trust management layer and another between the trust managers of individual nodes. The interfaces are as follows:

(1) Experience reports: Service engine → Trust manager
(2) Trust referrals: Trust manager ↔ Trust manager
(3) Policy updates: Trust manager → Service engine

Service usage is as normal, with the trust information just influencing protection mechanisms. The trust manager gathers usage experience and uses this together with the experience of collaborators to inform its trust in other nodes.

We then need to have a system to build trust as nodes gain experience of each other or learn of each other's reputation. In our system, trust can be updated in two ways:

1) Direct experience: Following a service usage event between two nodes, each node updates its trust in the other based on a measure of satisfaction with the event.

2) Reputation: Node A notifies other nodes in its *neighbourhood* of the trust score that it has for node B. This will change significantly following a security event.

How this neighbourhood is defined is significant. The neighbourhood of a node is the set of nodes with which it can communicate or with which it is willing to interact. The choice of neighbourhood nodes is up to each individual node to decide, though these collaborations will normally be two-way and based on topology (which in a wireless environment usually relates to location). In our simulations, we distribute nodes on a plane and define neighbourhood for each node based on distance, so nodes share information with nearby nodes.

The main benefit of this system is in using these trust scores to tune security settings. Trusted nodes can be dynamically provided with more privileges than untrusted nodes. In our system, as mentioned, we model all interaction between nodes in terms of services. Each node then sets a threshold trust level for access to each service it offers. If the trust score of a node decreases, for example due to detected suspicious activity by that node, the fewer services are made available to that node.

4 Design Model

4.1 Assumptions

We present a model here for using trust information to enhance security through collaboration in a rich multi-service decentralised environment. One of the problems of modelling decentralised systems like ad hoc networks is that it is unrealistic to take a top-down "bird's eye" view. Thus we model the set of nodes as those nodes of which a specific node *is aware*.

As mentioned earlier, our approach is service centric. All relevant activity between network nodes is modelled as service usage [16]. Nodes offer services and peer nodes use them. We assume that neighbour discovery and routing services are in place. We also assume that some kind of service registration and discovery is available to allow nodes to reach an understanding of the set of services available. Handorean & Roman [17] have proposed a secure service discovery technique for ad hoc networks.

It should be noted that the assumption of a common trust threshold across all nodes that provide the same service is something of a simplification. Even having all nodes share an understanding of the relative meaning of trust threshold values is non-trivial.

Some authors, such as Kamvar et al. [6], have proposed normalisation strategies for recommendations systems, in an attempt to address this.

We also assume authentication of identity, at least to the extent that a malicious node cannot masquerade as another more trusted node. This could be done, for example, by having nodes exchange public keys on their first interaction (in fact the public keys could be used as node identifiers). Further messages between those nodes could then be signed by the sending node's corresponding private key so that the recipient could at least be confident that the sender is the same entity as that for which it has been building up a trust profile. Verifying real identity is unimportant – in fact many entities may have the expectation of anonymity.

4.2 General Model

***Topology*:** Let $V_i = \{1,..., N_i\}$ be the set of nodes of which node i is aware. Some of these nodes will be *adjacent* to i, normally by reason of network topology. This can be modelled by an adjacency vector, $A_i = (a_{i,j})_{j=1,...,N_i}$, where $a_{i,j} = 1$ if pair (i, j) are neighbours and $a_{i,j} = 0$ otherwise.

Services: Let $S = \{S_1, \ldots, S_M\}$ be a set of services that may be provided. Each node j provides a set of services $S^j \subset S$. Some nodes will just be service consumers, so S^j will in those cases be empty.

***Trust threshold*:** Each service S_x has an associated *trust threshold* t_x, $0 \le t_x \le 1$.

***Representing Trust*:** We denote the local trust that node i has in node j as $T_{i,j}$. Each other node $k \in V_i$ will maintain its own local view of trust in j, which may, as we shall see, influence the local trust of node i in j. Note that $T_{i,j}$ is a trust *vector*. In the simplest case, this can just be a number, but such a number may have other related attributes, such as confidence in the trust score or recency, or the service(s) to which it relates.

***Trust initialisation*:** In the case where i has no prior knowledge of j, we will have $T_{i,j} = x$ where x is the *default trust*.

Trust decision: (using trust in service protection) When node j attempts to use service S_x provided by node i:

- Service use is permitted if $f_x(T_{i,j}) > t_x$, where the function f_x maps trust vector $T_{i,j}$ onto a scalar number in the range (0,1)
- Otherwise node j is blocked from using service S_x

***Trust update following service usage*:** After a service usage event, by node j on node i:

- positive outcome: $T_{i,j}$ is increased according to some algorithm
- negative outcome: $T_{i,j}$ is reduced according to some algorithm

In general, following service usage, we update $T_{i,j}$ according to:

$$T_{i,j} \leftarrow f_e(T_{i,j}E) \quad (1)$$

where E is a vector of attributes of the service usage event and f_e defines how trust is updated based on service usage experience.

***Trust update following referrals by a third party*:** Node i may receive a message from a third party node, k, indicating a level of trust node j. This can be modelled as node i adopting some of node k's trust level in node j, $T_{k,j}$. In general, following such a third party recommendation, we update $T_{i,j}$ according to:

$$T_{i,j} \leftarrow f_r(T_{i,j}, T_{i,k}, T_{k,j}) \quad (2)$$

where f_r is a function defining how trust is updated. This trust transitivity depends on $T_{i,k}$, as node i can be expected attach more weight to a referral from a highly trusted node.

4.3 Algorithms

Some algorithms for trust update. Both the way trust is updated based on service usage, and how referrals are issued and handled, have a profound impact on the usefulness of this kind of trust overlay system. The best algorithms will of course depend on network topology, node mobility, the range and types of service on offer, the attack threat model, and security requirements. The potential proportion of "bad guys" in the system (i.e. compromised nodes) has an impact, as does the scope for collusion between them. In this paper, we consider some candidate algorithms and how well they work in a selection of topology, security, and attacker collusion scenarios. How well they work is assessed in terms of system stability and improved resistance to attack.

Note that nodes are autonomous in our model and each might implement a different algorithm for updating and using trust. It can also be expected that a node's behaviour in terms of handling trust may change if it is hijacked.

The following are some examples of trust update algorithms:

Moving average: Each new assessment of trust is fed into a simple moving average, based on a sliding window. Direct experience can be given more weight than third party referrals in the averaging if desired. More advanced moving averages are also possible, where old data is "remembered" using data reduction and layered windowing techniques.

Exponential average: Each new assessment of trust is fed into a simple exponential average algorithm. Exponential averaging is a natural way to update trust as recent experience is given greater weight than old values, and no memory is required in the system, making it more attractive than using a moving average. Direct experience can be given more weight than referrals by using a different (higher) parameter. In our initial simulations, we update trust based on experience as follows:

$$T_{i,j} \leftarrow \alpha S + (1-\alpha)T_{i,j} \quad (3)$$

where parameter α, $0 \leq \alpha \leq 1$, defines the rate of adoption of trust based on experience. Note that having $\alpha = 0$ means that the trust value is unaffected by the experience. Having $\alpha = 1$ means that local trust is always defined by the latest experience and no memory is retained. In general, the higher the value of α, the greater the influence of recent service usage on the trust value maintained for that node. Lower values of α encourage stability of the system. In our initial simulations, we update trust based on referrals as follows:

$$T_{i,j} \leftarrow T_{i,j} - \beta T_{i,k} \left(T_{i,j} - T_{k,j} \right) \tag{4}$$

where β is a parameter indicating the level of influence that "recommender trust" has on local trust, $0 \leq \beta \leq 1$. Note that, the larger the value of $T_{i,k}$, (i.e. the more i trusts k), the greater the influence of k's trust in j on the newly updated value of $T_{i,j}$. Note that, if $T_{i,k} = 0$, this causes $T_{i,j}$ to be unchanged.

Exponential average, per service parameters: Parameters depend on service. Here, some services are more revealing than others of user's trustworthiness. Usage of a service that provides little opportunity for exploitation (say a service that allows read-only access to a system) shouldn't have much impact on trust in the user.

No forgiveness on bad experience: This in a draconian policy: a node behaving badly has its trust set to zero forever. This could be considered for critical services.

Second chance (more generally, n^{th} chance): Draconian policies are generally not a good idea. Intrusion detection and other security systems are prone to false alarms. Also, good nodes can be temporarily hijacked and should be given the opportunity to recover. Thus a suitable variation on the "no forgiveness" algorithms is to keep a count, perhaps over a sliding time window, of misdemeanours. Trust is set to zero, possibly forever, on n misdemeanours.

Hard to gain trust; easy to lose it: To discourage collusion between bad nodes, there is a case for making it hard to gain trust and easy to lose it. Thus attackers will require lots of effort to artificially increase their trust scores, either by repeated benign use of low-threshold services or by issuing repeated positive recommendations about one another. Even a little malicious activity will cause trust to fall significantly.

Use of corroboration: To prevent an attack by up to k colluding bad nodes, we could require positive recommendations from at least $k+1$ different nodes.

Some algorithms for collaboration. As already indicated, collaboration in our system is by the issuing of trust referrals. Certain rules are required to make this system workable in possibly large scale distributed environments. Some candidate approaches are as follows:

Random destination and frequency: At random intervals, a node i chooses a random other node that it knows about, j, to which to issue a referral. This approach ensures even node coverage over time, and scalability can be ensured by agreeing common upper bounds on the frequency of referrals.

On request: Node i only issues a referral to node j when asked for it. From node j's perspective, there is no guarantee that node i will respond.

On service usage experience: Node i only issues a referral relating to node k just after a service offered by node i has been used by node k. This means that the referral is based on fresh experience.

Only on reduction in trust: Referrals are only issued on detection of malicious activity. In this case, the trust system would act like part of a distributed intrusion detection system (IDS).

Each node specifies own trust receiving preferences: Here, each node specifies how often and under what conditions spontaneous trust reports are sent to it. A node may only wish to receive referrals if trust has changed significantly, for example.

5 Experiments, Results and Discussion

In this section, we present and discuss the results of some experiments to evaluate the stability of this system and its effectiveness in detecting attacks, including the incidence of false alarms. In particular we examine the following:

- Impact of having multiple services with different trust requirements.
- Influence of network topology and how a node's neighbourhood is defined.
- Security: attacker interaction with the reputation system, especially collusion between bad nodes to attempt to artificially raise their trust scores, and strategies to mitigate such collusion.

5.1 Multiple Services with Different Trust Requirements

We consider the situation where there are multiple services with different trust thresholds. We firstly just have three services (A, B, and C) so that we can more easily see the effects. We assign quite distinct trust thresholds to these: 0.2, 0.5 and 0.9, respectively.

We initially consider a 20-node fully connected network; i.e. every node has access to every other node to attempt service usage and to share trust referrals. In the next subsection, we repeat these experiments with a sparser network topology. A default initial trust score of 0.25 is chosen so that all nodes will have initial access to service A but will not have access to services B and C. In effect, nodes need to earn trust based on their usage of service A before they are allowed access to services B and C. To keep things straightforward, there are just two types of nodes. In Fig. 1 below, of the twenty nodes, two are 'bad' and the remaining eighteen are 'good'. We increase the number of bad nodes in Fig. 2.

We initially consider a fully connected network – i.e. every node has access to every other node to attempt service usage and to share trust referrals. In the next subsection, we repeat these experiments with a sparser network topology.

In the figures that follow, we plot the percentage of "good usage allowed" – the percentage of service usage attempts by good nodes that succeed. We also plot the percentage of "bad usage allowed" that is, the percentage of bad usage attempts that succeed. Ideally 100% of good usage attempts and 0% of bad ones should succeed.

To make the situation more realistic, we also model the fact that IDSs do not always get it right. Occasionally an attack goes unnoticed (false negative) or, alternatively, benign activity raises an alarm (false positive). We use simple Gaussian (normal) distributions to model the characteristics of service usage that are being monitored by the IDS. In the case of good nodes, a number is tagged onto each service usage using a normal distribution with mean 2.0 and standard deviation 2.0. Similarly, in the case of bad nodes, a number is attached to the service usage using a normal distribution with mean –2.0 and standard deviation 2.0. The sign of this number is then used by the pseudo-IDS to "detect" whether there has been an attack.

Figs. 1 and 2 show the percentages of good and bad usages allowed, over all three services, for a fully connected topology. The number of bad nodes is greater in Fig. 2.

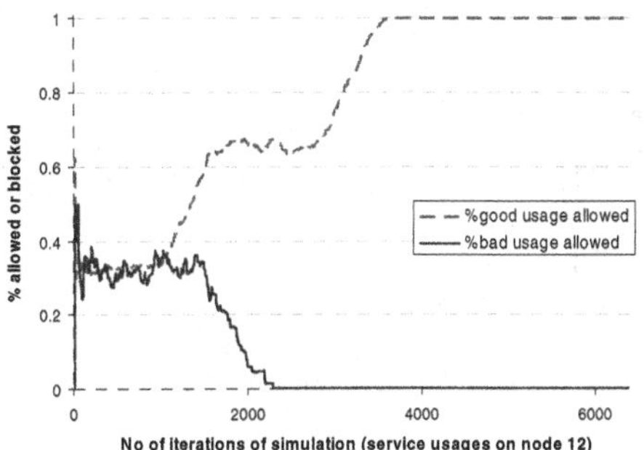

Fig. 1. % good usage allowed and % bad usage allowed, aggregated over all three services: 18 good nodes, 2 bad nodes, fully connected topology

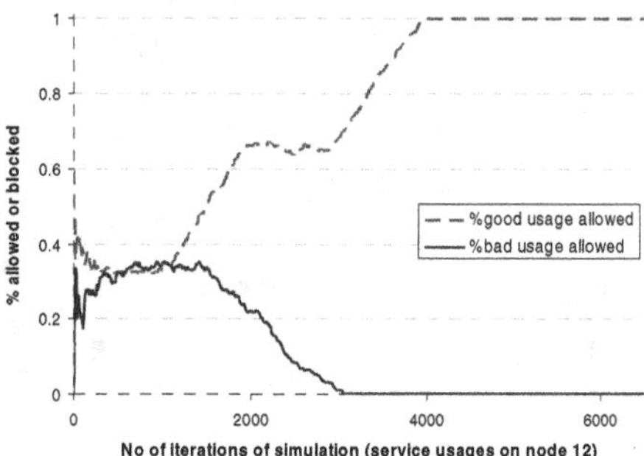

Fig. 2. % good usage allowed and % bad usage allowed, aggregated over all three services: 15 good nodes, 5 bad nodes, fully connected topology

In both these cases, the system works very well after trust has converged. When trust has been properly established, even with 25% of nodes behaving maliciously, all attempts by good nodes to gain access are allowed and the bad nodes are blocked. Note from the plateaus how the good nodes initially can mostly access just service A, then both services A and B, and finally all three services. Trust convergence takes a bit longer in the presence of bad nodes.

5.2 Influence of Network Topology and How a Node's Neighbourhood Is Defined

In the examples shown above, the neighbourhood of a node is defined as containing *every other* node (i.e. a fully connected network). Ad hoc networks are of course often less well connected and thus it is useful to study topological effects. Fig. 3 shows a randomly generated twenty node planar topology where connectedness depends on distance. Note that nodes 19 & 20 are 'bad' nodes in the simulations shown here. Furthermore, node 5 is effectively "hidden behind" a bad node.

Fig. 4 shows the effects of topology on service access adjudication in the multi service environment described in the previous section. The main impact of this relatively sparse topology is that some nodes are more significantly impacted by the behaviour of bad nodes. The success rate for access attempts by good nodes does not quite reach 100%, due mainly to one node (node 5 in our example) being "hidden" behind bad node 19.

A main effect of having a sparser network topology is that convergence is slower. This will vary from node to node, and is less of a problem for node 12 than some others, as node 12 is fairly well connected. The higher the proportion of bad nodes, the more potential there is for a poor community view to be maintained of poorly connected but well-behaved nodes. Despite good behaviour, trust in some good nodes might never get very high and they may be blocked from some services.

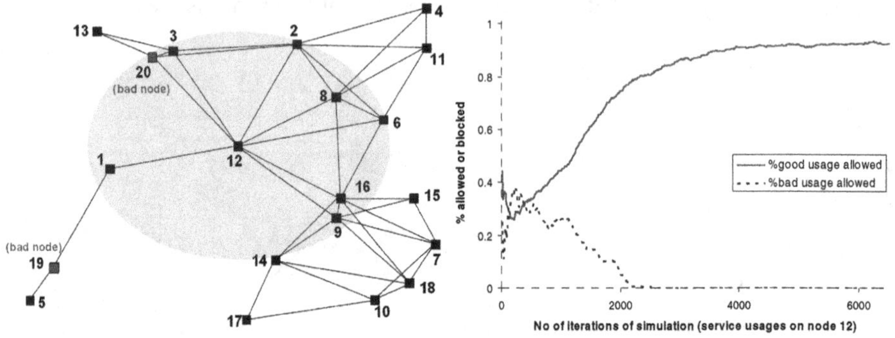

Fig. 3. Network topology used for simulations. The neighbourhood of node 12 is highlighted for illustration. Nodes 19 & 20 are bad nodes for some experiments.

Fig. 4. % good usage allowed and % bad usage allowed, aggregated over all three services: 18 good nodes, 2 bad nodes, topology as in Fig. 3

5.3 Anti-collusion Strategies

Attacker interaction with the system is of particular interest, especially collusion between bad nodes to attempt to artificially raise their trust scores. Fig. 5 compares the success rate of bad nodes in attempting to gain access to services when then act independently with when they act in collusion. Without collusion, each bad node, as well as behaving badly when granted service access, sends out low trust referrals about all nodes. We model collusion as where bad nodes are aware of each other and send out high trust referrals about each other and low ones about other (good) nodes. Not surprisingly, they are more effective at gaining access when actively colluding.

Other strategies are possible to attempt to break the system. One possibility is for a bad node to act benignly for a time in order to build up its reputation and gain access to more sensitive services than would otherwise have been possible. Strategies for countering this include (i) applying a referrer trust threshold and (ii) making it harder to gain trust than to lose it. Fig. 6 shows how such strategies can help. Each of the plots shown in this figure is for the proportion of bad usage allowed by the system in the face of colluding bad nodes as discussed above. With either application of a referrer trust threshold (of 0.5) or applying a factor (of 5) that makes gaining trust harder than losing trust, a significant improvement is achieved.

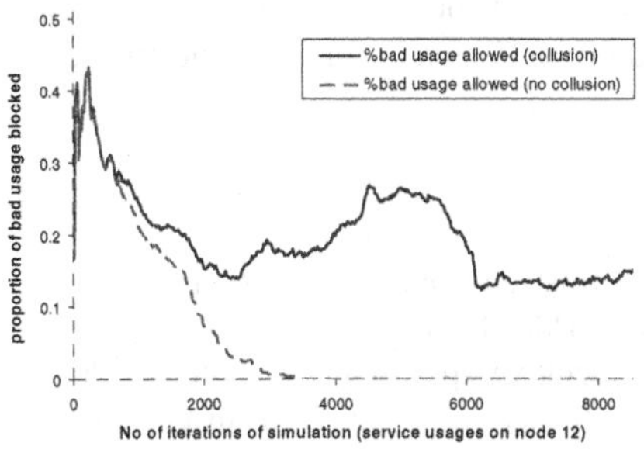

Fig. 5. % bad usage allowed where nodes actively collude compared with the same measure where nodes behave badly but do not work together. Aggregated over all three services: 15 good and 5 bad nodes.

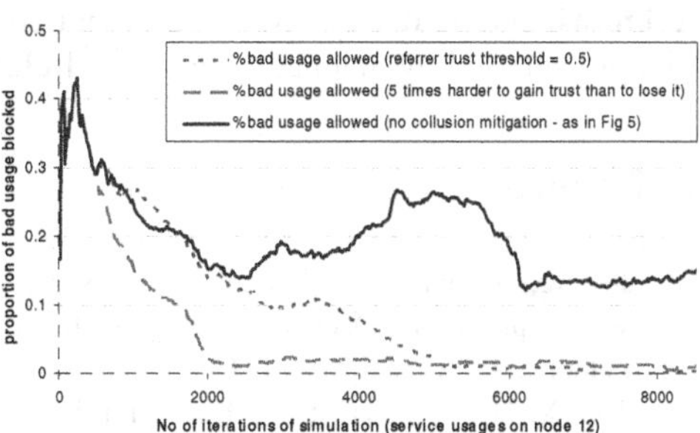

Fig. 6. % bad usage allowed where nodes actively collude. A comparison is shown between cases where there is no strategy to mitigate collusion and two anti-collusion approaches.

6 Conclusions and Future Work

We have described a collaborative approach to handling dynamic threats to services running on unstructured networks. In this, we have defined a dynamic system involving nodes, services and trust scores that appears to help to quickly and reliably isolate sources of attacks and restrict their access to the system. Services have associated trust thresholds – the more sensitive the service, the higher the threshold. New service users will initially just be able to access services with low thresholds, but repeated positive use of these can help the user gain sufficient trust to later access more sensitive services. The key to this kind of dynamic decentralised system is a stable means of updating trust scores that can learn from positive experiences and react to bad ones. Some inter-node services will be allowed and some will blocked by either end, depending on the trust score it has determined for the other. Simulation results presented in section 5 are encouraging.

Our simulations to date have just used relatively simple strategies for sharing and updating trust information. There is significant scope to refine the dynamics of the system. New algorithms need to be developed and evaluated for trust updates. A protocol is required to specify how and when trust information is shared between nodes. Further experiments are needed to explore the effects of how node neighbourhoods are defined.

A fuller security assessment of this system is required. Further possible collusion strategies of bad nodes need to be examined. It is also assumed in our model that authentication is provided by underlying services, but the implications of this assumption needs to be evaluated in some detail. Account also needs to be taken of privacy, data integrity and availability issues.

Further work is also required to assess the performance implications of having this kind of trust overlay. Though the system was designed with scalability in mind (the use of neighbourhoods, and a lack of any centralised services), various aspects such as choice of neighbourhood need to be optimised.

Acknowledgements. This work is supported by the European Commission (*OPAALS* FP6 project) and by Science Foundation Ireland (*Foundations of Autonomics* project).

References

1. Jøsang, A., Ismail, R., Boyd, C.: A survey of trust and reputation systems for online service provision. Decision Support Systems, 618–644 (March 2007)
2. Falcone, R., Castelfranchi, C.: Social trust: a cognitive approach. In: Castelfranchi, C., Tan, Y.-H. (eds.) Trust and Deception in Virtual Societies, pp. 55–90. Kluwer Academic Publishers, Dordrecht (2001)
3. Dingledine, R., Freedman, M., Molnar, D.: Accountability measures for peer-to-peer systems, Peer-to-Peer: Harnessing the Power of Disruptive Technologies, O'Reilly (2000)
4. Douceur, J.: The Sybil attack. In: Druschel, P., Kaashoek, M.F., Rowstron, A. (eds.) IPTPS 2002. LNCS, vol. 2429, pp. 251–260. Springer, Heidelberg (2002)
5. Gambetta, D.: Can we trust trust? In: Gambetta, D. (ed.) Trust: making and breaking cooperative relations, pp. 213–237. Basil Blackwell (1988)
6. Kamvar, S., Schlosser, M., Garcia-Molina, H.: The EigenTrust algorithm for reputation management in P2P networks. In: Proc. 12th World Wide Web Conf., Budapest (May 2003)
7. Gupta, M., Judge, P., Ammar, M.: A reputation system for peer-to-peer networks. In: Proc. 13th Int'l Workshop on Network and Operating Systems Support for Digital Audio and Video (NOSSDAV) (2003)
8. Damiani, E., De Capitani di Vimercati, S., Paraboschi, S., Samarati, P., Violante, F.: A reputation-based approach for choosing reliable resources in peer-to-peer networks, In: Proc. 9th ACM conference on Computer and Communications Security (ASIACCS) (November 2002)
9. Page, L., Brin, S., Motwani, R., Winograd, T.: The PageRank citation ranking: bringing order to the web, Technical report, Stanford Digital Library Technologies Project (1998)
10. Repantis, T., Kalogeraki, V.: Decentralized trust management for ad-hoc peer-to-peer networks. In: Proc. MPAC 2006, Melbourne (November 2006), http://www.smartlab.cis.strath.ac.uk/MPAC/
11. Zhou, L., Haas, Z.: Securing ad hoc networks. IEEE Network (November/December 1999)
12. Stajano, F., Anderson, R.: The resurrecting duckling: security issues for ubiquitous computing. IEEE Computer (Supplement on Security & Privacy), pp. 22-26 (April 2002)
13. Yang, H., Shu, J., Meng, X., Lu, S.: SCAN: self-organized network-layer security in mobile ad hoc networks, IEEE Journal of Selected Areas in Communications (February 2006)
14. Buchegger, S., Le Boudec, J.-Y.: A robust reputation system for mobile ad-hoc networks, EPFL IC Technical Report IC/2003/50, EPFL (July 2003)
15. McGibney, J., Botvich, D., Balasubramaniam, S.: A Combined Biologically and Socially Inspired Approach to Mitigating Ad Hoc Network Threats. In: Proc. 66th IEEE Vehicular Technology Conference (VTC), Baltimore (October 2007)
16. McGibney, J., Schmidt, N., Patel, A.: A service-centric model for intrusion detection in next-generation networks. Computer Standards & Interfaces, pp. 513-520 (June 2005)
17. Handorean, R., Roman, G.-C.: Secure service provision in ad hoc networks. In: Orlowska, M.E., Weerawarana, S., Papazoglou, M.M.P., Yang, J. (eds.) ICSOC 2003. LNCS, vol. 2910, pp. 367-383. Springer, Heidelberg (2003)

RAUU: Rate Adaptation for Unreliable Unicast Traffic in High Speed Networks*

Lihua Song[1], Haitao Wang[2], and Ming Chen[1]

[1] Institute of Command Automation,
[2] Institute of Communication Engineering
PLA Univ. of Sci. & Tech., 210007 Nanjing, China
{Minnehaha,Haitmail,Mingchen}@126.com

Abstract. While long term throughput not exceeding TCP with Reno congestion control algorithm is widely accepted as the criterion of weighing TCP friendliness, this may lead to resource waste in high speed networks due to Reno's known performance limits. Inspired by FAST TCP, a congestion control algorithm named Rate Adaptation for Unreliable Unicast traffic (RAUU) is proposed for unreliable unicast traffic in high speed networks to improve its efficiency while still holding friendliness to TCP. Being a rate-based approach to best fit unreliable unicast traffic, RAUU has made special design choices to alleviate the inherent contiguous loss problem of rate adaptation algorithms. Like FAST, it also tries to maintain appropriate number of extra packets in networks, and for that purpose it combines loss and delay as congestion signals. Theoretical analysis shows that in ideal networks RAUU has and will converge to its one and only equilibrium state where the number of extra packets is equal to the preset value. Plentiful simulation experiments confirmed that it could achieve similar performance to FAST as well as comparable throughput smoothness to TFRC while keeping TCP-friendliness at the same time.

Keywords: congestion control, unreliable unicast traffic, TCP-friendly, rate adaptation.

1 Introduction

Congestion control is one of the critical technologies in network service provision. Internet has been growing rapidly in the last two decades. Nevertheless, with the dramatically increasing size, speed, load, and connectivity, it continues to provide stable and efficient service without severe congestions. Most people believe that this is due to the congestion control mechanisms built in TCP and the fact that TCP contributes more than 90% traffic on the Internet. However, this situation may be changing now. With the growing popularity of streaming media applications, the ratio of non-TCP traffic to TCP traffic is increasing every year. The wide lack of a TCP-friendly congestion control mechanism in this kind of traffic will shake the foundation of stability and robustness of the Internet and bring potential congestion collapse risk to it.

* Supported by the National Natural Foundation of China under Grant No. 90304016.

D. Medhi et al. (Eds.): IPOM 2007, LNCS 4786, pp. 107–118, 2007.

To prevent non-TCP traffic from starving TCP, a lot of work has been done in the area of TCP-friendly congestion control algorithms. In those effort (represented by TFRC [1]), TCP-friendliness is usually weighted in the criterion of prevalent TCP congestion control algorithms. That is, a non-TCP flow is considered TCP-friendly if its long term throughput doesn't exceed that of a TCP flow using Reno or NewReno algorithm under the same conditions [2]. But the prevalent, or classic, TCP congestion control algorithms were designed for earlier networks and have many performance limitations in high-speed ones. In fact, improving TCP's stability, responsiveness and utilization in modern networks is exactly the subject of ongoing research on TCP congestion control today. If prevalent TCP congestion control algorithms are still used to restrict non-TCP traffic, not only network bandwidth will get wasted but also the congestion control research for non-TCP traffic will get behind TCP.

Recently, an end-to-end TCP congestion control algorithm named FAST [3-4] has become outstanding because of its distinguished performance in high speed networks. Inspired by FAST, we propose in this paper a rate adaptation algorithm named RAUU (Rate Adaptation for Unreliable Unicast traffic). Theoretical analysis as well as plentiful simulation experiments demonstrates that this algorithm does fit unreliable unicast traffic characteristics. In high speed network simulations, it achieved similar performance to FAST and smooth throughput comparable to TFRC. Meanwhile, it managed to maintain reasonable friendliness to TCP. The long run throughput of a RAUU flow was similar to that of FAST TCP under same conditions. And when contending with simultaneous TCP flows using NewReno congestion control algorithm, RAUU would not squeeze TCP's bandwidth share excessively.

2 Related Work

Work about TCP-friendly congestion control for unreliable unicast traffic can be classified according to the way they implement traffic regulation. There are two main methods a source can use to adjust the rate at which it injects packets into networks, namely window-based and rate-based methods.

In the window-based group, there are some approaches that simply duplicate TCP's congestion control mechanism into unreliable unicast traffic [5-7]. Just like TCP, they use congestion window to limit the number of outstanding packets and adapt to network state by adjusting the window size. Since TCP halves its window in response to a packet loss, these algorithms may produce abrupt throughput curves and significantly reduce the user-perceived quality. To alleviate this problem, some window-based approaches try to generalize TCP's Additive Increase/Multiplicative Decrease algorithm [8-9]. For example, while keeping throughput approximately the same, GAIMD [8] modifies TCP increase value and decrease ratio as parameters to smooth the throughput curve. But as we can see soon, the real effect is limited.

In the rate-based group, there are AIMD-like rate-based algorithms and equation-based algorithms. The former is rate-based variations of TCP's AIMD algorithm. They act similarly to TCP only except using rate regulation instead of window regulation [10-12]. So the inheritance of TCP's oscillating activity is not surprising. In contrast, equation-based approach, which is represented by TFRC [1], gives up AIMD algorithm and regulates sending rate according to a steady state TCP throughput equation [13]. By avoiding dramatic rate drops, TFRC achieves smooth output and satisfies the needs of real time applications. However, it is the Reno algorithm that the equation stands for. So as we argued before, this is not suit for high speed networks.

FAST is an end-to-end congestion avoidance algorithm that use delay in addition to loss as a congestion measure. FAST's control object is to maintain appropriate number of extra packets in the bottleneck buffers along the transfer path, to obtain both good utilization and quick responsiveness. Its packet-level congestion window update is based on the following iteration,

$$w \leftarrow (1-\gamma)w + \gamma\left(\frac{BaseRTT}{RTT} \cdot w + \alpha\right) \cdot \quad (1)$$

where $\gamma \in (0,1]$, w represents the congestion window size, *BaseRTT* is the round trip propagation delay, and α is the quantity of extra packets a flow attempts to maintain, which determines the protocol's fairness and convergence speed. FAST achieved much better utilization, convergence, stability and fairness than prevalent TCP congestion control algorithms in high speed network simulations [3-4].

3 The RAUU Algorithm

The Rate Adaptation for Unreliable Unicast traffic (RAUU) algorithm is designed to meet requirements from three aspects: first, being friendly to TCP; second, being efficient in high speed networks; third, regulating throughput smoothly and avoiding dramatic decrease. Inspired by FAST, RAUU also tries to maintain certain amount of extra packets in networks and in that process both loss and delay is its concern.

3.1 Rate-Based Approach

RAUU takes a rate-based approach because of the following considerations:

Window-based congestion control method is coupled with TCP's reliable transmission and flow control systems. In fact, sliding window system was initially designed for reliable data transfers. It ensures that packets would be delivered to the receiving application reliably and ordered even if the sending buffer and receiving buffer are limited in size. Congestion control is just a natural extension of that function by means of a congestion window.

Since RAUU's service target, i.e., the unreliable unicast traffic, doesn't require reliable or in order transmission, the need for a window system is not essential any more. On the contrary, a lot of trouble would be involved unnecessarily if we still use window-based approach like TCP. For example, accumulating ACK has little sense without reliable transmission. Now it is a problem about what kind of ACK should be used and how to guarantee an ACK's reliability. In [7], significant overhead is introduced due to the vector form of ACK and the acknowledgements sent for acknowledgements. Thus, window-based approach is an unnatural way to implement congestion control for unreliable traffic in following TCP. It should just regulate the rate directly. However, rate-based control has an inherent weakness. Without the protection provided by congestion window, a flow may have sent too many packets before it receives a congestion signal and thereby causes contiguous losses. Special treatment has been incorporated into the RAUU algorithm to alleviate this problem.

3.2 Elements

The basic idea of RAUU is to have the sender and receiver gather some performance parameters collaboratively, and then let the sender adjust its sending rate to make the number of extra packets conserved in networks quickly converge to a preset value. Specifically, the algorithm is composed of the following elements:

1. The sender gives a sequence number and a timestamp to each packet it send.
2. The receiver sends back an ACK to the sender every RTT given there are new data packets arriving. The content of an ACK includes average receiving rate and loss statistics during last RTT as well as timestamp of the most recent received packet.
3. On receiving an ACK, the sender computes the moving average round-trip-time *RTT* based on the current time and the timestamp returned, and then derive the queueing delay *qdelay* by subtracting round-trip propagation delay from it.
4. Based on whether there was packet loss in last RTT, the sender adjusts its sending rate accordingly:
   (a) If no packet lost,

$$rate_s = rate_r + \gamma(\alpha - rate_r \cdot qdelay)/2RTT \,. \tag{2}$$

   (b) Otherwise,

$$rate_s = \min(rate_s,\ rate_r,\ rate_r + \frac{\gamma(\alpha - rate_r \cdot qdelay)}{2RTT}) - \frac{1\,MPL}{RTT} \,. \tag{3}$$

Where $rate_s$ is the sending rate, $rate_r$ is the receiving rate fed back which represents the flow's current throughput. α is a preset number of extra packets. *MPL* is the maximum packet length. γ is a constant between 0 and 1.

5. For the new rate to take effect, the sender updates its sending rate every other RTT.
6. The sender halves its sending rate if no feedback is received over two RTTs.

The $\frac{BaseRTT}{RTT} \cdot w + \alpha$ part of equation (1) represents FAST's expected congestion window size, i.e., the size of congestion window when the number of extra packets reaches its preset value α. FAST adjusts its congestion window in proportion to the difference of expected window size and actual window size with ratio γ. If we do it the same way in RAUU, then the rate update equation will be $rate_s = (1-\gamma) \cdot rate_r + \gamma \cdot \alpha / qdelay$, because with queueing delay *qdelay* and α extra packets in networks the expected throughput will be $\alpha / qdelay$. However, RAUU is a rate-based approach. It doesn't have a congestion window to help it limit the amount of outstanding packets. As we noted before, if it sending too fast for the network to bear, there is a probability that so many packets have been sent before it receives the congestion notification that volumes of contiguous losses will occur and the network will experience oscillations. In fact, because the expected throughput is inverse proportional to the queueing delay, this risk will be very high when *qdelay* is small.

Following a different way, equation (2) regulates RAUU's sending rate in proportion to $(\alpha - rate_r \cdot qdelay)/2RTT$. Since there is a fixed part of propagation delay in RTT, besides α is constant, the increment of $rate_s$ in every control cycle will be limited. Also notice that $\alpha - rate_r \cdot qdelay$ represents the difference of preset and actual number of extra packets. Given $rate_r$ retains the same, one control cycle (two RTTs) later, this difference will shrink γ times according to equation (2). That is to say, the actual number of extra packets will steadily approach its preset value at the rate of γ times nearer per cycle, which complies with TCP FAST in essence.

When packet loss happened in the last RTT, equation (3) regulates the new sending rate as the minimum of old sending rate, receiving rate and would-be new sending rate provided no loss, less 1 *MPL/RTT*. Here approximate linear decrease is exploited instead of TCP's multiplicative decrease to provide some extent of smoothness for real time applications. Meanwhile in case of congestion exacerbation the sending rate will quickly decrease in response to $rate_r$ decrease and *qdelay* increase.

3.3 Refinements on Startup and Feedback Delay

In addition to above elements, further analysis and refinements have also been done on the startup phase and feedback delay problem.

According to equation (2), when the queueing delay is zero, the sending rate will increase constantly by sending $\gamma \cdot \alpha$ more packets every other RTT, which is unduly slow comparing to the idle environment at that time. For this reason, analogy to TCP, a "slow start" (actually it is quick) algorithm is created to replace equation (2) when *qdelay* = 0. However, due to RAUU's rate-based characteristic, this algorithm is some different from TCP's slow start algorithm. To prevent high sending rate from concussing network, the sending rate $rate_s$ is set to two times of $rate_r$ only in the first RTT of every control cycle during the startup phase (i.e., when *qdelay* = 0), and it goes back to safe $rate_r$ again in the second RTT. With this special treatment, the potential loss number is limited even if the sending rate exceeds available bandwidth because overload will now only last for one RTT.

When the sender receives an ACK and updates its sending rate according to equation (2) or (3), the *RTT* and *qdelay* samples it sees actually reflect network states about half RTT length time ago. And the network is changing persistently. To compensate this hysteresis, the difference of the current sending and receiving rate should be considered: $rate_s > rate_r$ indicates that the queueing delay keeps increasing and new sending rate should be somewhat smaller than the equation result; $rate_s < rate_r$ indicates that the queueing delay has decreased and new sending rate should be somewhat larger than the equation result. Based on this analysis, we use $\max(0,\ qdelay + \frac{rate_s - rate_r}{rate_r} \cdot \frac{RTT}{2})$ to replace *qdelay* in equation (2) and (3). Simulation experiments showed that this improvement expedites RAUU's convergence as well as alleviates its oscillation.

4 Equilibrium and Convergence

RAUU's property of equilibrium and convergence was analyzed theoretically. The result shows that with ideal network assumption, RAUU has and will converge to one and only equilibrium state where its extra packet amount equal to the preset number and its sending as well as receiving rate equal to its fair bandwidth share, i.e., an ideal state. Ideal network means network with synchronized and immediate feedback [14-15]. It is an assumption commonly used in analysis of congestion control algorithms.

Theorem 1. RAUU has one and only equilibrium state. In that state each RAUU flow maintains exactly as many as preset number (α) extra packets in networks.

Proof. One side, a RAUU flow is in equilibrium if its sending rate is equal to the receiving rate with no packet loss and the actual number of extra packets it maintaining is equal to the preset number α. Because given these conditions there will be $qdelay = \alpha / rate_r$, so the sending rate will remain unchanged in equation (2). Thus both receiving rate and number of extra packets will remain unchanged.

The other side, if a RAUU flow is in some equilibrium state, then according to the algorithm description there must be no packet loss (otherwise the sending rate will decrease) and meanwhile $rate_s = rate_r$ (or the queue is piling up or pouring down and the equilibrium will be broken). Now the algorithm degrades to equation (2) with $\gamma(\alpha - rate_r \cdot qdelay)/2RTT$ part equal to 0. And for the $\gamma(\alpha - rate_r \cdot qdelay)/2RTT$ part to be equal to 0 there must be $rate_r \cdot qdelay = \alpha$, which means that the flow is maintaining extra packets exactly amounting to α.

Combine both sides we get the RAUU having one and only equilibrium state conclusion. Since in equilibrium state each RAUU flow maintains number of extra packets as preset, all concurrent RAUU flows will share bandwidth fairly if they are all preset with the same α value. And this is true even for a mix of RAUU flows and FAST TCP flows.

Theorem 2. In ideal networks, starting from any state, RAUU will converge to equilibrium if there are no concurrent flows and router buffer's capacity is infinite.

Proof. Since there is only one flow, if we note bottleneck capacity as B and α/B as q_d, then we get $rate_s = rate_r = B$ and $qdelay = q_d$ in equilibrium state from theorem 1.

If the queueing delay *qdelay* is zero, then according to RAUU algorithm the sending rate will increase at a speed of nearly doubling every other RTT until it is not. So not losing generality, let's assume that initially there is $qdelay = q_d + \Delta q > 0$ and $rate_r = B$. Moreover, with the infinite buffer size and no feedback delay assumptions, the rate regulation of RAUU could be simplified to just equation (2). Taking in initial conditions we get

$$rate_s = B + \gamma \cdot (\alpha - B \cdot (q_d + \Delta q))/2RTT$$

Thus, two RTTs (i.e. a control cycle) later, the queueing delay would become

$$qdelay = \max(0,\ qdelay + (rate_s - B) \cdot 2\,RTT/B)$$

$$= \max(0,\ q_d + \Delta q - \gamma \cdot \Delta q)\ = q_d + (1-\gamma) \cdot \Delta q$$

Since $qdelay = q_d + (1-\gamma) \cdot \Delta q > 0$ (for $0 < \gamma < 1$), there would still be $rate_r = B$.

Go on like this, we can see that the queueing delay would become $q_d + (1-\gamma)^2 \cdot \Delta q$, $q_d + (1-\gamma)^3 \cdot \Delta q$, ……, and $q_d + (1-\gamma)^n \cdot \Delta q$ after 2, 3, …, and n control cycles respectively.

With $\lim_{n \to +\infty} (q_d + (1-\gamma)^n \cdot \Delta q) = q_d$, the queueing delay *qdelay* is actually approaching to q_d. Taking $qdelay = q_d$ and $rate_r = B$ into equation (2) we get $rate_s = B$. And that is to say, the RAUU flow will achieve its equilibrium state eventually.

Convergence of RAUU in real-condition networks with more than one concurrent flow was investigated by simulation experiments.

5 Simulation

By using network simulator ns-2 [16], we have carried out plentiful experiments to investigate RAUU's properties on many aspects such as convergence, throughput, delay, fairness, friendliness, and smoothness. We describe part of these results below.

5.1 Performance

In order to examine RAUU performance under various network conditions, a group of experiments were carried out in single as well as various multi-bottleneck network simulation topologies. Here is the result of single-bottleneck topology. This was a "dumbbell" network (figure 1) with bottleneck capacity of 800 Mbps and buffer size of 2,000 packets. 4 RAUU flows, whose round trip propagation delay are 200 ms, 100 ms, 150 ms and 250 ms respectively, one by one joined in the contention for bottleneck bandwidth at 400 seconds intervals. They then quitted in inverse turn. The simulation packet length was 1,500 Bytes. The extra packet parameter α was set to 200. $\gamma = 0.5$. (Experiments below employed same parameter settings.) Curves of bottleneck throughput, bottleneck queueing delay, bottleneck packet loss, as well as curves of throughput and packet loss for each flow, and the fairness index [14] over all flows are shown in figure 2.

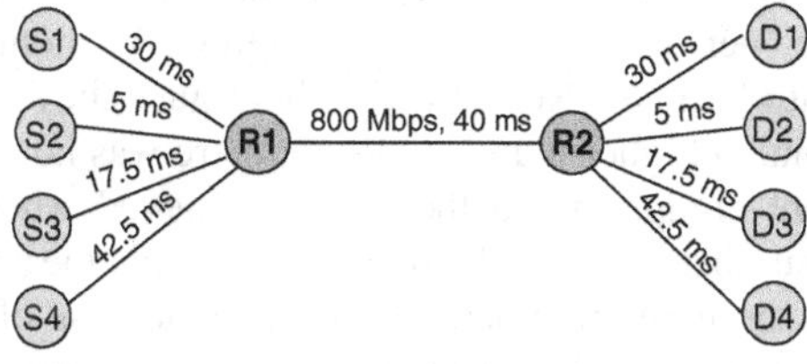

Fig. 1. Single bottleneck simulation topology

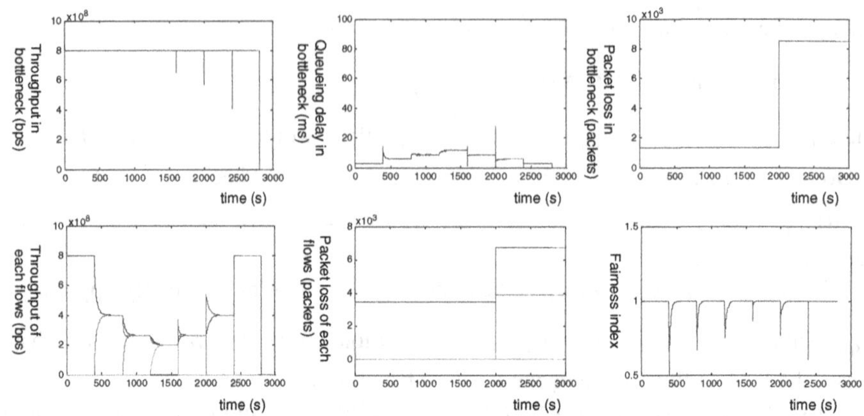

Fig. 2. RAUU performance in single bottleneck network simulation

From figure 2 we see that RAUU behaved very well. The bottleneck's bandwidth got fully utilized. The queueing delay was quite small. After a flow entering or leaving, each flow could converge to equilibrium and enjoy its fair share. The fairness index over all flows was 1 most of the time. All of these indicate that RAUU has good bandwidth allocation property. In fact, RAUU's behavior resembled that of FAST under same conditions [3], except the packet loss and relative slow convergence due to smoothness requirement. Nevertheless, all loss events happened at the time when the first flow started and the third flow quitted, and the number of lost packets was about a transmission amount in half RTT, which means that RAUU effectively controlled the contiguous loss problem of rate-based approaches. Results of two-bottleneck and multi-bottleneck topologies are similar, which are omitted here.

5.2 TCP-Friendliness

Friendliness of RAUU to TCP was investigated from two aspects. Impact of RAUU flows on TCP flows using prevalent congestion control algorithm NewReno was examined first. Then contention between RAUU and FAST TCP was studied.

Use the 800 Mbps single bottleneck network again. 10 flows, all started randomly in 1 second to contend for bottleneck bandwidth, had the same round trip propagation delay of 100 ms. The number of NewReno TCP flows varied from 0 to 10, the rest were all RAUU flows. RAUU's settings were like before. The average throughput of all NewReno flows over 100 seconds simulation time and the one of all RAUU flows over the same time period are shown in figure 3, along with the average retransmission amount of all NewReno flows. This figure shows that although RAUU achieved throughput much higher than NewReno, there was no significant decrease in NewReno's throughput comparing to the one it achieved when it contending with itself (i.e., when the number of RAUU flows was 0). This means that RAUU did not squeeze NewReno's fair share intemperately, it just made use of the spare bandwidth NewReno unable to use in high speed networks. The retransmission amount of NewReno flows was not affected evidently by RAUU flows either.

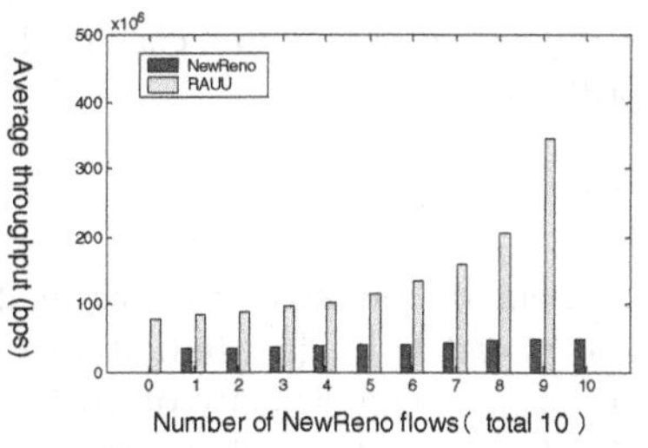

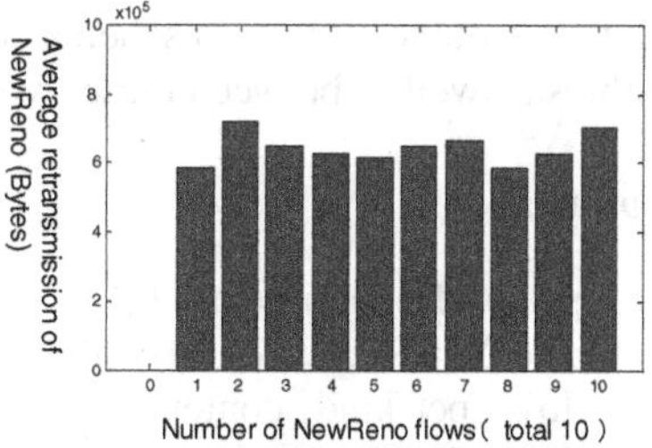

Fig. 3. RAUU contending with NewReno TCP

In the second part of this group of experiments, equal number of FAST TCP flows and RAUU flows (the total number is *2n*, where *n* is a positive integer) started in turn to contend for a bottleneck link. All FAST and RAUU flows had the same round trip propagation delay of 180 ms; their extra packet parameter α was set to 200 packets. Other settings were like before. The surface of FAST average throughput over the last 100 seconds simulation time when the bottleneck bandwidth *B* varied from 100 Mbps to 900 Mbps and the number *n* varied from 1 to 32 is shown in figure 4. Here the throughput values have been normalized to FAST's fair bandwidth share. So value 1 represents perfect fairness. Note that bottleneck utilization was always close to 100%. Figure 4 also illustrates contention details between RAUU and FAST when *B*=600 Mps and *n*=8. The upper right picture shows for each flow the throughput curves over time. The lower right picture gives the curve of system's fairness index over time.

We can see from figure 4 that the throughput of FAST TCP was close to its fair share. Since both RAUU and FAST try to maintain the same number of extra packets in networks, two flows each using one of these algorithms should achieve the same throughput in theory. We can also see that with new flows joining in, both existing RAUU flows and FAST flows could converge to the equilibrium state quickly. In equilibrium state the system's fairness index was stabilized around value 1, which

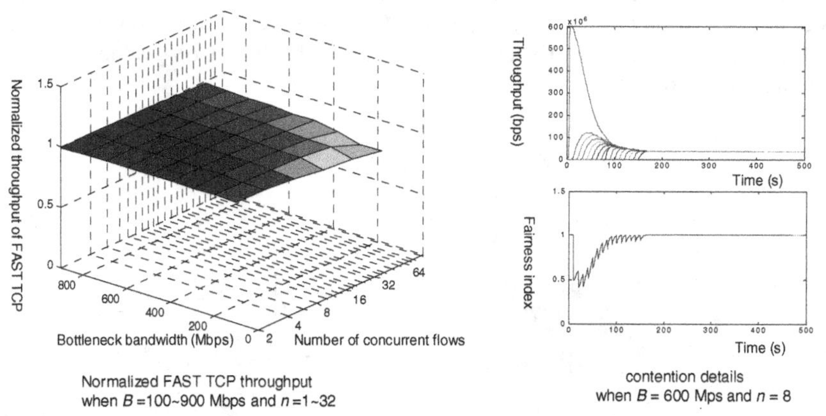

Fig. 4. RAUU contending with FAST TCP

means that the resource allocation scheme achieved good fairness within each of these two algorithms as well as between these two algorithms.

5.3 Smoothness

In the third group of experiments, RAUU's throughput smoothness was investigated and compared to NewReno, GAIMD (0.31, 0.125) [8] and TFRC [1]. Initially, we let 16 flows (4 flows per kind) contend for a bottleneck link with capacity of 16 Mbps. Selecting 2 flows from each kind non-purposely, figure 5 shows their throughput evolvement averaged over 200 ms units. What can be told from this figure is that the 4 congestion control algorithms we seeing about clearly categorized into two groups: TFRC and RAUU managed to achieve smoothly throughput regulation in contrast to the fluctuation of NewReno and GAIMD.

Throughput's smoothness property is closely related to the time scale over which the throughput is computed. To quantitatively evaluate and compare the smoothness of different congestion control algorithms on various time scales, we made use of the method employed in [1], in which a flow's throughput sequence from time t_0 to time t_1 averaged over time units of length δ is regarded as a statistic sample of size n (n=(t_1 - t_0)/δ) and the sample's CoV (Coefficient of Variation) attribute is taken as a measure of the smoothness of this flow on time scale δ. The smaller CoV obtained, the smoother throughput is.

Figure 6 depicts the throughput CoV attribute for these four algorithms under different network conditions. The CoV value was computed on various time scales ranging from 0.2 s to 12.8 s. While there were small shifts in absolute values and all algorithms' smoothness became better on larger time scales, it is not hard to see that the CoV attribute of RAUU and TFRC was better than that of NewReno and GAIMD under same network condition and on same time scale at all times. This confirmed the result of figure 5. Since the number of concurrent flows was kept low, the above experiments were actually carried out in light-loaded, congestion-free environments where bottleneck loss rate never exceeded 1%. By adding ON/OFF cross flows, we also carried out the same experiments in slight congestion (with loss rate about 2%), moderate congestion (with loss rate about 15%), and severe congestion (with loss rate about 50%) environments. Except difference in absolute CoV values, the result resembled figure 6.

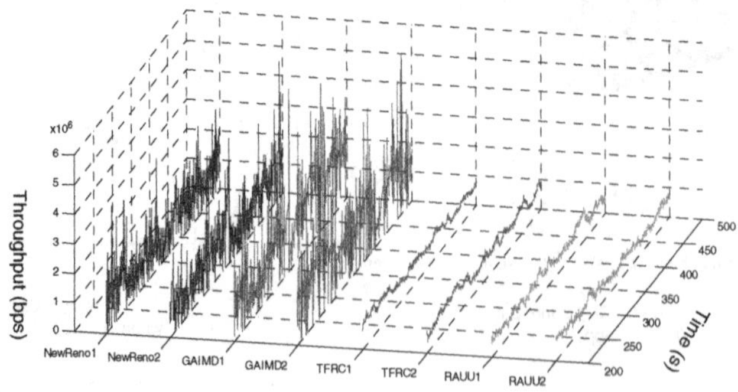

Fig. 5. Throughput evolvement of RAUU, NewReno, GAIMD, and TFRC over time

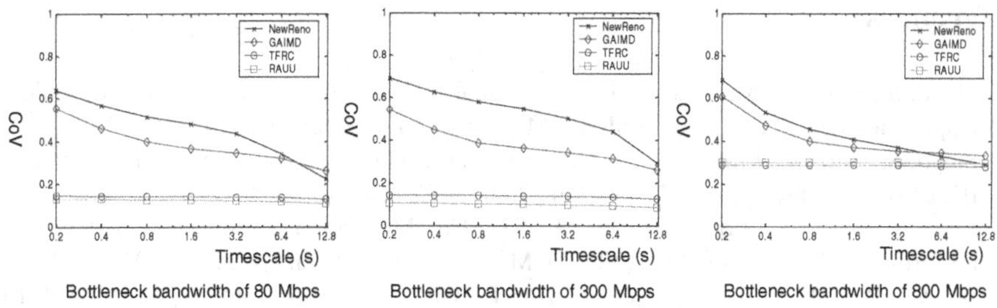

Fig. 6. Quantified smoothness attributes of RAUU, NewReno, GAIMD and TFRC on various time scales in different network environments

Smoothness of TFRC is well known. Now the above experiments indicate that RAUU could achieve similar smooth throughput while being much less complex.

6 Conclusion

Prevalent TCP congestion control algorithms such as Reno, NewReno are usually used as criterion in evaluating TCP-friendliness of congestion control algorithms for non-TCP traffic. However, with their well-known performance problems, this will result in inefficiency and resource waste in high speed networks.

Illuminated by FAST TCP, a rate adaptation algorithm, RAUU, is proposed for unreliable unicast traffic to improve efficiency in high speed network environments while still being friendly to TCP. Considering the distinct characteristics of unreliable unicast traffic, this algorithm employs a rate-based approach and has made special treatment in its design to avoid the inherent contiguous loss problem of rate-based approaches. Like FAST, RAUU takes maintaining appropriate number of extra packets in networks as its control object. It combines congestion signals of loss and delay, and adjusts its sending rate according to the difference of preset and actual number of extra packets. Theoretical analysis shows that in ideal networks RAUU has and will converge to the equilibrium state where the number of extra packets equal to the preset value. Plentiful simulation experiments exhibit its good behavior in high speed networks. It achieved similar performance to FAST as well as comparable throughput smoothness to TFRC while keeping TCP-friendliness at the same time. Neither RAUU squeezes NewReno's bandwidth share excessively, nor does it acquire more throughput over long time than FAST in same conditions.

The preset number of extra packets (i.e., parameter α) has significant impact on network performance because all flows are trying to keep that number of extra packets in bottleneck buffers and the number of concurrent flows is varying all the time. Resolvent of FAST is to set α statically by consulting link capacity. But this is incapable of keeping up with the fast moving contention level on networks. In [17] we suggest a potential scheme in which network measurement technologies are employed to select proper α value dynamically based on the current network state.

References

1. Floyd, S., Handley, M., Padhye, J., et al.: Equation-Based Congestion Control for Unicast Applications. In: Proc. of ACM SIGCOMM, Stockholm, Sweden, pp. 43–56 (2000)
2. Floyd, S., Fall, K.: Promoting the Use of End-to-End Congestion Control in the Internet. IEEE/ACM Transactions on Networking 7, 458–472 (1999)
3. Jin, C., Wei, D.X., Low, S., et al.: FAST TCP: Motivation, Architecture, Algorithms, Performance. In: Proc. of IEEE INFOCOM, HongKong, China, pp. 2490–2501 (2004)
4. Jin, C., Wei, D.X., Low, S., et al.: FAST TCP: From Theory to Experiments. IEEE Network 19, 4–11 (2005)
5. Rhee, I., Ozdemir, V., Yi, Y.: TEAR: TCP Emulation at Receivers – Flow Control for Multimedia Streaming. Technical Report (2000), http://www.csc.ncsu.edu/faculty/rhee/export /tear-page/techreport/tearf.pdf
6. Kohler, E., Handley, M., Floyd, S.: Datagram Congestion Control Protocol (DCCP). Internet Draft (2005), http://www.ietf.org/internet-drafts/draft-ietf-dccp-spec-13.txt
7. Floyd, S., Kohler, E.: Profile for DCCP Congestion Control ID 2: TCP-like Congestion Control. Internet Draft (2005), http://www.ietf.org/internet-drafts/draft-ietf-dccp-ccid2-10.txt
8. Yang, Y.R., Lam, S.S.: General AIMD Congestion Control. In: Proc. of the 8th IEEE International Conference on Network Protocols, Osaka, Japan, pp. 187–198 (2000)
9. Bansal, D., Balakrishnan, H.: Binomial Congestion Control Algorithms. In: Proc. of IEEE INFOCOM, Alaska, USA, pp. 631–640 (2001)
10. Rejaie, R., Handley, M., Estrin, D.: RAP: An End-to-End Rate-Based Congestion Control Mechanism for Realtime Streams in the Internet. In: Proc. of IEEE INFOCOM, New York, USA, pp. 1337–1345 (1999)
11. Sisalem, D., Schulzrinne, H.: The Loss-Delay Based Adjustment Algorithm: A TCP-Friendly Adaptation Scheme. In: Proc. of Workshop on Network and Operating System Support for Digital Audio and Video, pp. 215–226. Cambridge,UK (1998)
12. Sisalem, D., Wolisz, A.: LDA+: A TCP-Friendly Adaptation Scheme for Multimedia Communication. In: Proc. of IEEE International Conference on Multimedia and Expo (III). GMD-Fokus, Germany, pp. 1619–1622 (2000)
13. Padhye, J., Firoiu, V., Towsley, D., et al.: Modeling TCP Throughput: A Simple Model and its Empirical Validation. In: Proc. of ACM SIGCOMM. Vancouver, California, pp. 303-314 (1998)
14. Chiu, D., Jain, R.: Analysis of the Increase and Decrease Algorithms for Congestion Avoidance in Computer Networks. Computer Networks & ISDN Systems 17, 1–14 (1989)
15. Loguinov, D., Radha, H.: End-to-End Rate-Based Congestion Control: Convergence Properties and Scalability Analysis. IEEE/ACM Transactions on Networking 11, 564–577 (2003)
16. The Network Simulator – ns-2 (2006), http://www.isi.edu/nsnam/ns/
17. Song, L.H., Wang, H.T., Chen, M.: Congestion Control Scheme Aiming at P2P Applications in High-Speed Networks. In: Proc. of the 19th International Teletraffic Congress, Beijing, China, pp. 2099–2108 (2005)

Multi-source Video Streaming Suite

Pablo Rodríguez-Bocca[1,2], Gerardo Rubino[2], and Luis Stábile[2]

[1] Instituto de Computación - Facultad de Ingeniería
Universidad de la República,
Julio Herrera y Reissig 565, 11300,
Montevideo, Uruguay
prbocca@fing.edu.uy
[2] Inria/Irisa
Campus universitaire de Beaulieu
35042 Rennes, France

Abstract. This paper presents a method for the distribution of video flows through several paths of an IP network. We call the approach *multi-source* focusing on the fact that, from the receiver's point of view, the stream is obtained from several different sources. A typical use of our procedure is in the design of a P2P network for streaming applications. We implemented a tool in VLC which can send video streaming from multiple sources. In this paper, we describe the method together with the design decisions and the operation characteristics of the tool. Finally we consider, at the receiver side, the perceived quality, automatically measured using the PSQA methodology. We illustrate the work of the tool by means of some tests.

Keywords: multi-source, Internet Protocol, VLC, PSQA.

1 Introduction

The main objective of our tool is the distribution of video flows through several paths of an IP network. We call the approach *multi-source* focusing on the fact that, from the receiver's point of view, the stream is obtained from several different sources. The idea is to resist to the possible failures of the servers distributing somehow the associated risks. Therefore, multi-source streaming techniques are used typically in P2P networks, where the peers become often disconnected.

Previous studies (see [1]) show that in 30-80% of the cases there is at least a better end-to-end path, from the quality perspective, for a transmission. In the particular case of video flows, it is then easier to recover from isolated losses than from consecutive ones [2]. Another advantage of this approach is that it helps to balance the load on the networks (we are not going to address this issue in this paper). The basic difference between the multi-path and multi-source concepts is that in the former case we consider a single source sending data to the destination following different paths. The main difficulty here is the routing aspects: we must specify and control which path must be followed by which packet in the flow. The multi-source approach implies that there are multiple and independent sources for the signal, and that some general scheme allows

D. Medhi et al. (Eds.): IPOM 2007, LNCS 4786, pp. 119–130, 2007.

the receiver to get the stream from a set of servers. Since they are different, the path that the packets will follow would be a priori different, without any need of acting on the routing processes. Of course, in practice, the different paths could share some nodes, which are good candidates to become bottlenecks of the communication system. Detecting such cases is an active research area [3].

A few studies consider the implementation needed to split and to merge stream content, especially when it is live video streaming. For instance, Nguyen and Zakhor [4] propose to stream video from multiple sources concurrently, thereby exploiting path diversity and increasing tolerance to packet loss. Rodrigues and Biersack [5] show that parallel download of a large file from multiple replicated servers achieves significantly shorter download times. Apostolopoulos [1,6] originally proposed using striped video and Multiple Description Coding (MDC) to exploit path diversity for increased robustness to packet loss. They propose building an overlay composed of relays, and having each stripe delivered to the client using a different source. Besides the academic studies, some commercial networks for video distribution are available. Focusing in the P2P world (our main application here), the most successful ones are PPlive [7], SopCast [8], and TVUnetwork [9].

The paper is organized as follows. Section 2 introduces all relevant concepts for our project. In Section 3, a conceptual architecture of the components and the complete solution is presented. It consists of modules added to the open code VLC project of VideoLAN [10]. In Section 4 some first experimental results are introduced. The main contributions of this work and conclusions are in Section 5.

2 Multi-source Streaming

We will focus on the case of a P2P architecture where the video flow is decomposed into pieces and sent through the network. Each node receives the flow from different sources and builds the original stream before it is played. At the same time, it will, in general, send each of the substreams to a different client, acting then as a server. At the beginning, the (single) initial stream is split into several substreams, according to some specific scheme.

With these goals, we need the two basic following services: (i) a flexible "splitter" where the original stream is decomposed into different substreams, and (ii) a module capable of reconstructing the original stream from the set of substreams, or a degraded version of it if some of the substreams had losses, or is simply missing. Associated with these services we want to design a very flexible transport scheme, allowing to transport the original information plus some redundancy, with high control on which server sends which part of the flow and of the redundant information (see below). We also want to be able to measure, at the receiver side, the perceived quality, using the PSQA methodology.

Let us now briefly describe these different components of the project. The splitter must be able to obey a general scheme where we decide how many substreams are to be used, and which of the frames are to be sent through which of the substreams. This generality allows us to send, for instance, most of the frames through the highest performing nodes in the network, or to balance the load among the different components, or to adopt a scheme where the type of frame is used to take the routing decision. It must also be possible to send an extra fraction r of the original stream, or of a specific population of the frames (for instance, the I frames) again according to any pre-specified scheme. If $r = 0$ there is no redundancy at all. If $r = 0.2$ we send 20% of supplementary redundant data, and $r = 1$ means that the original stream is actually sent twice to the receiver(s) We not consider $r > 1$ because, in a real system, this would mean too much bandwidth consumption. At the client side, we must be able to reconstruct the original stream if we receive all the frames, but also if only a part of the stream arrives; this typically happens when some of the servers fails (that is, it disconnects from the P2P network). Last, the client must be able to identify the lost frames and to compute statistics from them. We want also that the client computes the value of the quality of the received stream, as evaluated by the PSQA tool, for instance for auditing the quality of the transmission system, or for controlling purposes.

2.1 The PSQA Technology [11]

PSQA stands for Pseudo-Subjective Quality Assessment. It is a technique that allow the building of a function mapping some *measurable* parameters chosen by the users into a numerical assessment of the perceived quality. The measures are taken on the client side, and a good example is the end-to-end loss rate of frames. The mapping is done by showing many sequences to real human users, where these parameters took many different values, and performing subjective tests in order to evaluate their quality values. Then, after statistically filtering the results provided by these tests, a statistical learning tool is used that allows to build the PSQA function. To use PSQA on the client side, we must be able to measure the parameters chosen at the beginning of the process, and then to call the function. The latter is always a rational function of the parameters, whose value can be obtained very quickly, allowing PSQA to operate, if necessary, in real time. The parameters are chosen at the beginning of the process. They can be qualitative or quantitative. Their choice depend on the network and/or the application at hand, and of course the resulting PSQA procedure will be better if the a priori choice is as relevant as possible.

We know from previous work on PSQA that the loss process is the most important global factor for quality. In this paper, we consider the loss rates of video frames, denoted by LR, and the mean size of loss bursts, $MLBS$, that is, the average length of a sequence of consecutive lost frames not contained in a longer such sequence. See [11] and the references therein for more details about PSQAs.

3 Design

This section will describe the main design decisions. There are K servers and each server is identified by an index between 1 and K. See Figure 1 for an illustration.

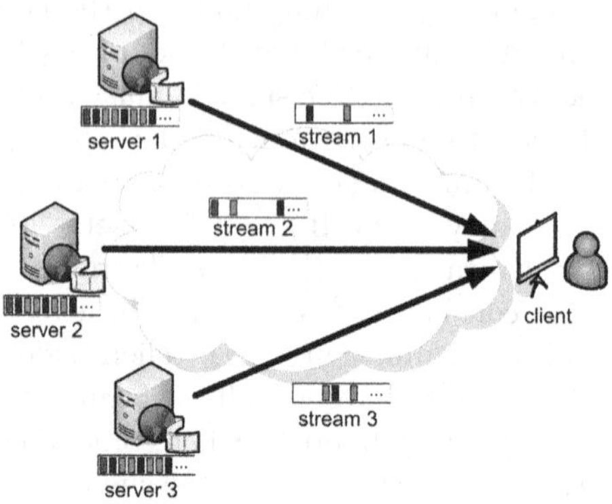

Fig. 1. The multi-source scheme

3.1 Global Architecture

The implementation consists of three VLC modules, one at the server side and two at the client side. They are called `msource-server`, `msource-bridge` and `msource-client` (see Figure 2).

The module `msource-server`, located at the server, decides which frames are to be sent to the client. It builds a substream for audio and another one for video. The basic constraint to satisfy is, of course, that each frame in the original stream must be sent at least once by one of the servers. Once the frames selected by `msource-server`, the module `standard` of VLC sends them to the client.

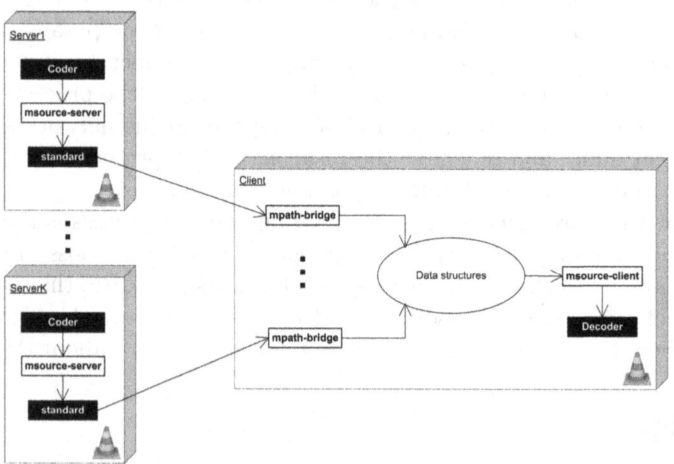

Fig. 2. Architecture multiple-source in VLC

At the client's side, there are K modules `msource-bridge`, one per server and each with its own buffer. The kth one receives the frames sent by server k. The last module, `msource-client`, receives the flows sent by the K modules `msource-bridge` and reconstructs the stream. This task consists of ordering the frames according to their decoding time (called DTS). The output of a module `msource-client` can be stored on the local disk, sent somewhere through the network (both tasks are performed by the module `standard`, which is a part of the VLC package) or played by the client.

The ideal case is when all the servers start their transmissions simultaneously and keep synchronized, but this never happens in practice. It means that the system must handle the possible shift between the servers. Moreover, since we don't assume that the servers are identical, we implement a strategy allowing to control the bandwidth used by each of them.

3.2 Basic Server Algorithm

The strategy used by the servers allows to control the bandwidth used by each of them, for instance, to be able to send more information through the most reliable one, or through the one having the best performance. This control is implemented by acting on the percentage of frames of each type that the server must send. We must decide which fraction $p_k^{(c)}$ of the class c frames, $c \in \{\text{I, P, B, A}\}$ [1] is going to be send through server k. We must then have

$$\sum_{k=1}^{K} p_k^{(c)} = 1, \quad c \in \{\text{I, P, B, A}\}. \tag{1}$$

The sending algorithm works as follows. All the servers run the same pseudo-random number generator and build a sequence of pseudo-random reals uniformly distributed on $[0, 1]$. Not only they run the same generator but they use the same seed. In this way, they obtain the same sequence of real numbers $(u_1, u_2, \cdots)$ behaving as a realization of the corresponding sequence of i.i.d. random variables uniformly distributed on $[0, 1]$.

Now, look at $(p_k^{(c)})_{k=1,\cdots,K}$, for fixed c, as a probability distribution, and call $X^{(c)}$ the corresponding random variable. It can be sampled from an uniform random variable U using the classical algorithm given by

$$X^{(c)} = k \iff P_{k-1}^{(c)} \leq U < P_k^{(c)}, \tag{2}$$

where $P_j^{(c)} = p_1^{(c)} + \cdots + p_j^{(c)}$, $j = 1, \cdots, K$, and $P_0^{(c)} = 0$.

Now, let f_n be the nth frame of the stream, $n \geq 1$, and let $c_n \in \{\text{I, P, B, A}\}$ be the class of f_n. Any of the K servers will then run the same algorithm. They all sample the random variable $X^{(c)}$, and obtain the same value s_n, with $1 \leq s_n \leq K$. That is, they all look at the type of f_n, and use (2). Server s_n

[1] In MPEG, basically there are three classes of video frames (I, P and B), and one class of audio frames (A).

sends f_n and the remaining servers don't. This construction guarantees that one and only one server sends each frame, and since s_n behaves as a realization of random variable $X^{(c)}$, after sending many frames the fraction of type c frames sent by server k will be close to $p_k^{(c)}$. Observe that this method allows to control the bandwidth of each server in a scalable way (there is no limit on the number of servers nor on the distribution of the load among them).

3.3 Controlling the Redundancy Level

With our method we can send some redundancy to the client. This, together with our splitting procedure, adds robustness to the system, in order to face the problem of possible server failures (recall that we call failure any event causing the server to stop sending, for instance, because it simply left the network). As we will see in next subsection, redundancy also allows us to design a simple solution to the problem of synchronizing. We describe here the way we control the redundancy level in the system.

We allow the system to send redundant frames up to sending again the whole stream, and we provide a precise mechanism to control with high precision the redundancy level and the distribution of the supplementary load among the K servers. For this purpose, we implement a system where a given frame is sent either once or twice, and in the latter case, by two different servers. Our method allows to control the redundancy level per class, for instance, for each class of frames (I, P, B). We denote by $r^{(c)}$ the fraction of supplementary class-c frames that will be sent to the client (by the set of servers). So, each server k must decide if it sends frame f_n as the "original" frame, as a copy for redundancy, or not at all. The procedure described before allows to choose the server that sends the frame as the original one. For the redundacy, the implementation is also probabilistic. We proceed as follows. Together with the stream of pseudo-random numbers used to make the first assignment we described before, the set of servers build a second sequence $(v_1, v_2, \cdots)$ with the same characteristics (the same for all servers, the same seed). Suppose that frame f_n is of class c and that it is assigned to some other server j, $j \neq k$. Then, using the second sequence $(v_1, v_2, \cdots)$, server k samples a second random variable with values in the set $\{1, 2, \cdots, K\}$ to decide at the same time if a redundant copy of the frame is to be send and if it must send it. Let us call $Y_j^{(c)}$ this random variable. Its distribution is the following: $\Pr(Y_j^{(c)} = j) = 1 - r^{(c)}$ and if $m \neq j$,

$$\Pr(Y_j^{(c)} = m) = \frac{p_m^{(c)}}{\sum_{h:\, h \neq m} p_h^{(c)}} r^{(c)} = \frac{p_m^{(c)}}{1 - p_j^{(c)}} r^{(c)}.$$

If $Y_j^{(c)} = j$, no redundancy is sent. If $Y_j^{(c)} = m$, $m \neq j$, server m is the only one to send the frame as a redundant one.

3.4 Client Behavior

The client must reconstruct the stream using the flows received from the different servers. It will work even if some of the servers are missing (failed). Each server sends its streams marked with a time stamp indicating its playtime with respect to a specific reference (in VLC, 1/1/1970). Assuming all the servers synchronized, the algorithm is simple: it consists of selecting as the next frame the one having the smallest value of playtime. The problem is the possible shift between the servers (whose locations are different in the network). This can be due to the conditions encountered by the packets in their travelling through the network, or by the processing of the frames at the servers themselves. The key idea for knowing this shift in the transmission time of the servers consists of using the redundant frames.

First, let us look at the identification of the redundant frames. Each frame is sent with a time stamp corresponding to its playtime at the client side (computed by VLC). When receiving it, a Message-Digest algorithm 5 (MD5) [12] is computed and a dictionary is maintained to see if an arriving frame has already been received (necessarily by a different server). If the frame arrives for the first time, we store its MD5 together with its time stamp. Assume now that the same frame arrived from two different servers i and j, and call τ_i and τ_j the corresponding time stamps. The difference between these values is the (current) shift between the two servers. We denote it by Δ_{ij}, that is, $\Delta_{ij} = \tau_i - \tau_j$. Let us denote by Δ the $K \times K$ matrix (Δ_{ij}) where we define $\Delta_{ii} = 0$ for all server i. Observe that $\Delta_{ji} = -\Delta_{ij}$. Following these observations, we maintain such a matrix, initialized to 0, and we modify it each time a new redundant frame is detected (actually we do it less often, but this is a detail here). Observe that rows i and j in the matrix, $i \neq j$, are obtained by adding a constant element per element (or, equivalently, see that $\Delta_{ij} = \Delta_{ik} + \Delta_{kj}$). This constant is precisely the delay between the corresponding servers. The same happens with the columns. All this in other words: if we receive $K-1$ redundant pairs corresponding to $K-1$ different pairs of servers, we can build the whole matrix.

Each time we update matrix Δ, we can compute the index d of the most delayed server (if any), by choosing any row[2] in Δ and by computing its smallest element. Then, using for instance row 1, $d = \text{argmin}\{j : \Delta_{1j}\}$. When the client now looks for the next frame to choose, it scans the K buffers corresponding to the K servers. Let us denote by τ_k the time stamp of the first frame in the kth buffer (assume for the moment that no buffer is empty). Then, we first make the correcting operation $\tau'_k = \tau_k + \Delta_{dk}$, that is, we synchronize with the time of the most delayed server, and then we look for the server m where $m = \text{argmin}\{k : \tau'_k\}$. Next frame to be played will be the one in head of buffer m. This works as long as there are frames in buffer d. If it is empty after a play, we must wait for an arrival there, because we have no information about which server will be sending next frame to play. Of course, we must wait until some time out because if server d for some reason stopped sending, we block the system if we remain waiting for its frames. After some amount of time, we move to the frame having the smallest time to play using the previous procedure.

[2] Actually, we can choose any row if all the entries of Δ have been computed; otherwise, we choose the row having the largest number of computed entries.

Observe that, at the price of an extra computation at the client's side, we are able to synchronize efficiently the substreams without any signalling traffic, and using the same data that protects the system against failures.

4 Evaluation and First Results

For testing the correctness of our prototype, we implemented a program that collects complete traces of the sequences of frames sent by the servers and received by the client. These traces allows us to determine, for each frame, which server sent it and if it was played of not by the client. The program also collects some other data as the amount of frames sent, the used bandwidth, etc.

4.1 Testing the Bandwidth Used

The goal of the tests we will describe here is to measure the bandwidth used by the servers. This can be compared with the values provided by a theoretical analysis of the system, to check the consistency of the implementation. We will use two different models for distributing the load among the servers. The first one consists of sharing it uniformly among the K servers, that is, all of them send the same fraction $1/K$ (quantitatively speaking) of the global stream. In the second model, we use a geometric distribution: server i sends $1/2^i$th of the stream, for $i = 1, 2, \cdots, K-1$ and server K sends the fraction $1/2^{K-1}$.

Consider the uniform case, in which we send the stream with a redundancy level of r. If $BW_{K,i}^{unif}$ denotes the bandwidth used by server i, then clearly $BW_{K,i}^{unif} = (1+r)/K$. The case of our geometric load is a little bit more involved. If $BW_{K,i}^{geo}$ is the bandwidth used by server i in this case, we have

$$BW_{K,i}^{geo} = \begin{cases} \dfrac{1}{2^i} + \dfrac{r}{2^i}\left(1 - \dfrac{1}{2^i}\right) & \text{if } i \neq K \\ \dfrac{1}{2^{K-1}} + \dfrac{r}{2^{K-1}}\left(1 - \dfrac{1}{2^{K-1}}\right) & \text{if } i = K \end{cases}, \quad K \geq i \geq 1. \tag{3}$$

In our tests, we used the value $r = 0.5$. In Table 1 we show the Mean Squared Error between the bandwidth measured during our experiments and the theoretical values in the two distribution models considered, uniform and geometric. We evaluated the used bandwidth by extracting information automatically computed by the standard VLC modules. This consists of multiplying the number of frames sent in each class (I, P, ...) by the average size of these frames, and then by dividing by the used time. We then get an approximation of the used bandwidth. This, plus the losses (see next subsection) explains the differences between expected and observed values. Observe that we sum over all servers, which means that the effects of the randomness used are not responsible of any part of the observed differences.

Table 1. Mean Squared Error between theoretical and observed values of the used bandwidth, as a function of the total number K of servers. The error is computed summing on i, the server index, from 1 to K.

K	Uniform	Geometric
1	0.00000	0.00000
2	0.00005	0.00005
3	0.00013	0.00322
4	0.00006	0.00310
5	0.00008	0.00243
6	0.00005	0.00207
7	0.00006	0.00171
8	0.00004	0.00149
9	0.00009	0.00134
10	0.00011	0.00125

Table 2. Estimated loss rates after synchronization, as a function of the number K of servers used, for the two load models considered

K	loss rate (uniform load)	loss rate (geometric load)
1	0.0000	0.0000
2	0.0049	0.0049
3	0.0070	0.0080
4	0.0083	0.0066
5	0.0080	0.0070
6	0.0080	0.0072
7	0.0083	0.0220
8	0.0093	0.0186
9	0.0090	0.0182
10	0.0129	0.0222

4.2 Measuring Losses

In our preliminary prototype, there are some frame losses at the beginning of the transmission, because we assume no specific effort is made to synchronize the K servers (this extreme situation is considered for testing purposes). There is, then, a transient phase during which the system will loose information until there have been enough redundancies allowing to synchronize the flows using the procedure described before. Then, during the transmission, in other stressing experimental situations, some of the servers may have very few frames to send, which can make the synchronization process again slower to converge, until redundant frames are finally sent.

We computed the loss rates by comparing the sent and received frame sequences. We arbitrarily eliminated the first parts of the flows until observing 50

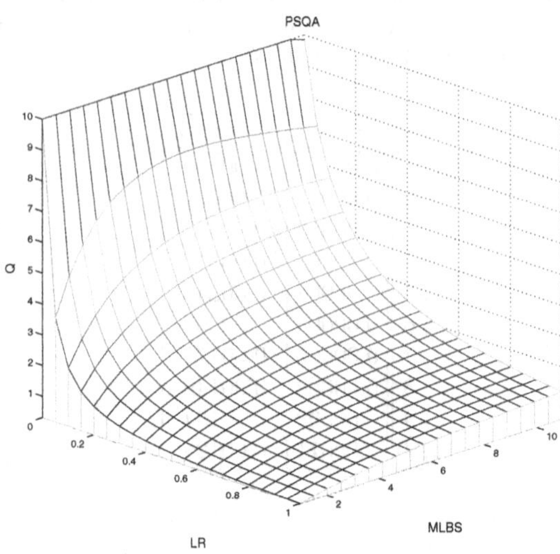

Fig. 3. The PSQA function used

consecutive frames correctly sent, because this implies in general that the K servers are synchronized. Again, the goal is to check that even using this rough testing procedure, the observed loss ratios are small enough. We used the same loads as in Subsection 4.1. In Table 2 we show the observed loss rates, during $2K$ experiments using k servers, $1 \leq k \leq K$, and both the uniform and geometric load distributions.

Observe the fact that in the geometric model, we measure some "peak" values of the loss rates for high values of K. This is due to the fact that in this load model there are servers that send a small number of frames. Specifically, all the servers with index $i \geq 7$ send $(1/2^6)$th of the original video. These servers never succeed in synchronizing because of the lack of redundant frames sent. This implies that the decoding time of the frames sent by these servers will not be correctly computed because the shifts will not been known by the system. The final consequence of this is that those frames will be discarded.

4.3 Perceived Quality

In this last subsection, we consider the perceived quality of the reconstructed stream. In the experiments mentioned here, the goal is to test the consistency of the system, and not its performances (this will be the object of further studies). However, we underline here the way the connection is made with the quality as perceived by the user.

The PSQA technology provides us with a function $Q = Q(LR, MLBS)$. In Figure 3 we plot the curve obtained following the PSQA procedure (see [11] for a general presentation of the PSQA approach and metrics such as LR and MLBS; see [13] for an example of application). First, observe that the quality is much more sensitive to losses than to the average size of the loss bursts. If we want to estimate the quality as perceived by the user, we must just measure the *LR* and *MLBS* and call the Q function using the measured values as input data. For instance, consider the loss rates obtained using the uniform load model, as shown in Table 2. Figure 3 shows that the worst case is when MLBS = 1. Using this value and the loss rates given in the first column of Table 2, we obtain, as perceived quality estimates, values going from 10 (maximal quality) to approximately 9, which is almost maximal. For the geometric load (somehow an extreme situation), we observe loss ratios up to 0.0222, which in the worst case of MLBS = 1, translates into a quality level of about 5 (this corresponds to the characterization "fair" in the subjective testing area). The reason was already commented: the servers having little to send have no time to synchronize. This is telling us that the losses due to the synchronization problem we observed, even if the situation which generated them is a little bit too pessimistic, should be diminished. This is commented in the last concluding section.

5 Conclusions

This work shows that a multiple source method for distributing live video can be efficiently implemented with a fine control on its redundant capacities. We also show that we can connect such a transport system with a tool allowing to measure the perceived quality as seen by the end user (the client), which can be used for controlling purposes.

One of the results we obtained is that the redundancy of our system not only provides robustness against servers' failures (typically, due to nodes leaving the networks) but also allows to implement a method for synchronizing the flows without any other signalling process.

The first measures we did show also that the present prototype needs some improvements to reduce the losses due to the possible lack of synchronization (even if they are not very important in number). In order to diminish these losses, we are currently working on adding extra redundancy for a better synchronization (for instance, by sending at the beginning, all the frames by all the servers, during some pre-specified time or until a pre-chosen number of frames have been sent). Also, when a server has almost no frame to send, we also can force it to send redundant data, again for allowing the client to estimate the drifts with accuracy enough.

One of the tasks needed for further work is to explore the performances of our methods in different failure scenarios. The goal of this study will be to explore the appropriate values of the number K of servers to use, together with the best redundancy level r. After such a performance evaluation analysis, the search for optimal tuning parameters can be considered.

References

1. Apostolopoulos, J., Trott, M.: Path diversity for enhanced media streaming. IEEE Communications Magazine 42(8), 80–87 (2004)
2. Apostolopoulos, J.: Reliable video communication over lossy packet networks using multiple state encoding and path diversity. In: VCIP 2001. Proc. Visual Communication and Image Processing, pp. 392–409 (2001)
3. Rubenstein, D., Kurose, J., Towsley, D.: Detecting shared congestion of flows via end-to-end measurement. IEEE Trans. on Networking 3 (2003)
4. Nguyen, T.P., Zakhor, A.: Distributed video streaming over Internet. In: Kienzle, M.G., Shenoy, P.J. (eds.) Proc. SPIE, Multimedia Computing and Networking 2002. Presented at the Society of Photo-Optical Instrumentation Engineers (SPIE) Conference, vol. 4673, pp. 186–195 (2001)
5. Rodriguez, P., Biersack, E.W.: Dynamic parallel access to replicated content in the internet. IEEE/ACM Trans. Netw. 10(4), 455–465 (2002)
6. Apostolopoulos, J., Wee, S.: Unbalanced multiple description video communication using path diversity. In: IEEE International Conference on Image Processing, pp. 966–969. IEEE Computer Society Press, Los Alamitos (2001)
7. PPLive Home page (2007), `http://www.pplive.com`
8. SopCast - Free P2P internet TV (2007), `http://www.sopcast.org`
9. TVUnetworks home page (2007), `http://tvunetworks.com/`
10. VideoLan home page (2007), `http://www.videolan.org`
11. Rubino, G.: Quantifying the Quality of Audio and Video Transmissions over the Internet: the PSQA Approach. In: Barria, J. (ed.) Design and Operations of Communication Networks: A Review of Wired and Wireless Modelling and Management Challenges, Imperial College Press (2005)
12. IETF Network Working Group: The MD5 Message-Digest Algorithm (RFC 1321) (1992)
13. Bonnin, J.-M., Rubino, G., Varela, M.: Controlling Multimedia QoS in the Future Home Network Using the PSQA Metric. The Computer Journal 2(49) (2006)

QoE Monitoring Platform for Video Delivery Networks

Daniel De Vera[2], Pablo Rodríguez-Bocca[1,2], and Gerardo Rubino[2]

[1] Instituto de Computación - Facultad de Ingeniería
Universidad de la República,
Julio Herrera y Reissig 565, 11300,
Montevideo, Uruguay
prbocca@fing.edu.uy
[2] Inria/Irisa
Campus universitaire de Beaulieu
35042 Rennes, France

Abstract. This paper presents a full video delivery network monitoring suite. Our monitoring tool offers a new view of a video delivery network, a view based on the quality perceived by final users. We measure, in real time and automatically, the perceived quality at the client side by means of the recently proposed PSQA technology. Moreover, we improve PSQA's efficiency and robustness for video analysis by studying the flows at the frame level, instead of the packet level previously considered in the literature. The developed monitoring suite is a completely free-software application, based on well-known technologies such as the Simple Network Management Protocol or the Round Robin Databases, which can be executed in various operating systems. In this paper we explain the tool implementation and we present some of the first experimental measurements performed with it.

1 Introduction

As a consequence of the opening of video content producers to new business models, the more bandwidth availability on the access network (on the Internet, cellular networks, private IP networks, etc.) and the explosion in the development of new hardware capable of reproducing and receiving video streams, the area of Video Delivery Networks is growing nowadays at increasing speed.

A common challenge of any VDN deployment is the necessity of ensuring that the service provides the minimum quality expected by the users. Quality of Experience (QoE) is the overall performance of a system from the users' perspective. QoE is basically a subjective measure of end-to-end performance at the services level, from the point of view of the users. As such, it is also an indicator of how well the system meets its targets [1].

To identify factors playing an important role on QoE, some specific quantitative aspects must be considered. For video delivery services, the most important

D. Medhi et al. (Eds.): IPOM 2007, LNCS 4786, pp. 131–142, 2007.

one is the perceptual video quality measure. Accurate video-quality measurement and monitoring is today an important requirement of industry. The service provider needs to monitor the performance of various network layers and service layer elements, including those in the video head-end equipment (such as encoders and streaming servers) as well as at the home network (such as the home gateway and STB). Some solutions are being delivered by niche vendors with a specific focus in this area [2,3], by large telecommunication infrastructure providers as part of an end-to-end VDN solution [4, 5, 6], or by a fully in-house development.

Monitoring tools can be classified into two different categories: active and passive. An active monitoring tool sends traffic through the network for performing its measurements. A passive one uses devices to watch the traffic as it passes through the measuring points. This paper describes a platform architecture belonging to the class of the active monitoring tools, that use probe nodes distributed in the network, with a centralized data collector using based on [7]. The traditional way to monitor video quality [2,3] in a VDN is a manual process, where experts observe the quality continuously in some displays located logically in different stages of the network (typically in the output of the encoding process and in a simulated client situated in the head-end of the network, where the experts are). In a IP network with losses and congestion, it is necessary to be in the edge of the network to measure accurately the quality, but this is not possible because the perceived quality measure is actually a manual process. To avoid that, the usual approach to assess the performance of a VDN is to use a well chosen metric, that we know plays an important role in quality, such as the loss rate of packets, or of frames, and to analyze it in the system of interest. In this paper we instead address the problem of directly measuring *perceived* quality in by means of the Pseudo Subjective Quality Assessment (PSQA) technology [8,9]. PSQA is a general procedure that allows the automatic measure of the perceived quality, accurately and in real-time. Moreover, we extend the technique and improve its efficiency for video analysis by studying the flows at the frame level, instead of the packet level previously considered in the literature.

The paper is organized as follows. Section 2 introduces the measurement methodology used in our platform. In Section 3, the architecture of the platform is described. In Section 4, we report on some experimental results allowing to illustrate the platform use. The main contributions of this work are then summarized in Section 5.

2 Methodological Considerations

Before describing our methodology, recall that in the most important specifications for video, MPEG-2 and MPEG-4, the transmission units are the *frames*, which are of three main types: the Intra frames (I), the Predicted frames (P) and the Bidirectional or Interpolated frames (B). An I frame codes an image, a P frame codes an image based on a previously coded I or P frame (by coding the differences, based on motion compensation) and a B frame is also a predicted one based on past as well as future P or I frames. The frame sequence between two consecutive I frames is a group of pictures (GOP) in MPEG-2 and a group of video object planes (GVOP) in MPEG-4. MPEG-2 and MPEG-4 also share a common concept called *user data*, which corresponds to byte sequences pertaining to an user application that can be inserted inside a stream. This can be done in many places, at the different abstraction levels defined in the specifications. The GOP header is the lowest one (this means that between two consecutive I frames we will find at most one piece of user data). As we will see in the following sections, the user data concept will be a fundamental piece in our audit platform design.

Quality Measurements. Let us consider here the different ways of evaluating the perceived quality in a video delivering system. Perceived video quality is, by definition, a subjective concept. The mechanism used for assessing it is called *subjective testing*. It consists of building a panel with real human subjects, which will evaluate a series of short video sequences according to their own personal view about quality. An alternative is to use a (smaller) panel of experts. In the first case, we will get the quality of the sequences as seen by an average observer. In the second case, we can have a more pessimistic (or optimistic, if useful) evaluation. The output of these tests is typically given as a Mean Opinion Score (MOS). Obviously, these tests are very time-consuming and expensive in manpower, which makes them hard to repeat often. And, of course, they cannot be a part of an automatic process (for example, for analyzing a live stream in real time, for controlling purposes). There exist standard methods for conducting subjective video quality evaluations, such as the ITU-R BT.500-11 [10]. Some variants included in the standard are: Double Stimulus Impairment Scale (DSIS), Double Stimulus Continuous Quality Scale (DSCQS), Single Stimulus (SS), Single Stimulus Continuous Quality Evaluation (SSCQE), Stimulus Comparison Adjectival Categorical Judgement (SCACJ) and Simultaneous Double Stimulus for Continuous Evaluation (SDSCE). The differences between them are minimal and mainly depend on the particular application considered. They concern, for instance, the fact that in the tests the observer is shown pre-evaluated sequences for reference (which in turn, can be explicit or hidden), the quality evaluation scale (and the fact that it is discret or continuous), the sequence length (usually around ten seconds), the number of videos per trial (once, twice in succession or twice simultaneously), the possibility to change the previously given values or not, the number of observers per display, the kind of display, etc.

Other solutions, called *objective tests*, have been proposed. Objective tests are algorithms and formulas that measure, in a certain way, the quality of a stream. The most commonly used objective measures for video are: Peek signal to noise ratio (PSNR), ITS' Video Quality Metric (VQM) [11], EPFL's Moving Picture Quality Metric (MPQM), Color Moving Picture Quality Metric (CMPQM) [12], and Normalization Video Fidelity Metric (NVFM) [12]. With a few exceptions, objective metrics propose different ways of comparing the received sample with the original one, typically by computing a sort of distance between both signals. So, it is not possible to use them in an real-time test environment, because the received and the original video are needed at the same time in the same place. But the most important problem with these quality metrics is that they often provide assessments that do not correlate well with human perception, and thus their use as a replacement of subjective tests is limited.

Pseudo Subjective Quality Assessment (PSQA). In [8] a hybrid approach between subjective and objective evaluation has been proposed. It is a technique allowing to approximate the value obtained from a subjective test but automatically. The idea is to have several distorted samples evaluated subjectively, that is, by a panel of human observers, and then to use the results of this evaluation to train a specific learning tool (in PSQA the best results come from the Random Neural Networks one [13]) in order to capture the relation between the parameters that cause the distortion and the perceived quality.

Let us briefly describe the way PSQA works. We start by choosing the parameters we think will have an impact on quality. This depends on the application considered, the type of network, etc. Then, we must build a testbed allowing us to send a video sequence while freely controlling simultaneously the whole set of chosen parameters. This can be a non-trivial task, especially if we use a fine representation of the loss process. We then choose some representative video sequences (again, depending on the type of network and application), and we send them using the testbed, by changing the values of the different selected parameters. We obtain many copies of each original sequence, each associated with a combination of values for the parameters, obviously with variable quality. The received sequences must be evaluated by a panel of human observers. Each human observer evaluates many sequences and each sequence is evaluated by all the observers (as specified by an appropriate test subjective norm). After this subjective evaluation, we must perform a statistical filtering process to this evaluation data, to detect (and eliminate, if necessary) the bad observers in the panel (a bad observer is defined as being in strong disagreement with the majority). All these concepts have well defined statistical meanings. At that stage enters the training process, which allows learning the mapping $\nu()$ from the values of the set of parameters into perceived quality. After the training phase, PSQA is very easy to use: we need to evaluate the values of the chosen parameters at time t (that is, to measure them), and then to put them into the function $\nu()$ to obtain the *instantaneous* perceived quality at t.

In our work, we focused on two specific parameters concerning losses, because we know from previous work on PSQA that the loss process is the most important global factor for quality. We consider the loss rate of video frames, denoted by *LR*, and the mean size of loss bursts, *MLBS*, that is, the average length of a sequence of consecutive lost frames not contained in a longer such sequence. The *MLBS* parameter captures the way losses are distributed in the flow. It is important to observe that in previous work using the PSQA technology the analysis was done at the packet level. Here, we are looking at a finer scale, the frame one. While packet-level parameters are easier to handle (in the testbed and from the measuring point of view in the network), frame-level ones provide a more accurate view of the perceived quality.

An Extensible Measurement Framework. We distinguish two main components within our measuring system: a set of video players and a family of centralized monitoring servers. The video players are an improved version of the VideoLan client software (VLC) [14]. Each VLC client performs the measuring tasks, and makes the measures available to the servers using the Simple Network Management Protocol (SNMP) [7]. A collector server polls the clients to obtain the SNMP MIB values of the measured parameters. Concerning SNMP, our application uses the SNMPv3 [7,15] version and our designed MIB module is compliant with the SMIv2 [16,17,18].

The parameters measured in the extended VLC client come from two different data sources: dynamically calculated information (e.g., video bitrate, I-Frame mean size, P-Frame mean size, B-Frame mean size, codecs detection and so on) and information included within the own stream. As mentioned before, the user data defined in MPEG-2 and MPEG-4 allows to insert application's information inside the stream. The measurement framework defines rules about how to tag a stream (for instance, what information should be inserted in the user data, how it should be formatted, and where it should be placed). The inserted information is captured and used by the extended VLC client the parameters calculation. This flexibility allows our tool to evolve smoothly, adapting to changes in the network, the applications, or the users' behavior, by simply modifying the input parameters used to build the quality evaluation metric.

In Table 1 we show the current parameters measured within our framework. The *user data* contains the number of I, P and B frames from the beginning of the stream. We send it at the beginning of each GOP. The extended VCL players count the frames received per class and, when the new user data arrives, compare them to their own counters. This way, frame losses are detected with high precision.

Table 1. Measurement framework parameters

Frames Related Information
Losses per frame type
Frames expected to receive per frame type
Frames received per frame type
Mean size per frame type
Frames bitrate
Streams Related Information
Streaming server IP and port
Transport protocol
Container format
Video and audio codecs
Clients Related Information
Client active streams quantity
Time since the beginning of a stream execution

An important fact to consider is that the overhead added in the stream by the user data insertion is completely negligible (in our tests: 19 bytes in 22700 bytes in MPEG-2 and 19 bytes in 281700 bytes in MPEG-4).

3 The Audit Platform

Architecture. Inside the VDN we want to monitor, there are five basic components (Fig. 1): the streaming server, the probes, the data collector server, the PSQA Learning Tool and the Webstat application. The streaming server streams the video's content over the VDN. Probes are VLC players strategically located in the network, taking specific measures and sending reports using SNMP. The data collector server polls each probe of the VDN (using SNMP) in order to gather the probes's reports. The PSQA Module is where the perceptual quality value is computed. Finally `Webstat` provides a web interface for the probes's reports presentation.

Streaming Server. The function of the streaming server is to provide multimedia content to the VDN. This content can be coded using different video specifications (MPEG-2, MPEG-4...), audio specifications (MP3, FLAC...), container formats (MPEG-TS, MPEG-PS, OGM...) and it can be streamed over different transport protocols (HTTP, UDP, RTP, MMS...). As we mentioned before, the measurement framework requires the user data insertion in the streamed content over the VDN. For this purpose, we have two different alternatives: inserting the user data in the streaming server on the fly, thus using a specific server, or inserting them in a post-encoding process, thus without any need for a specific

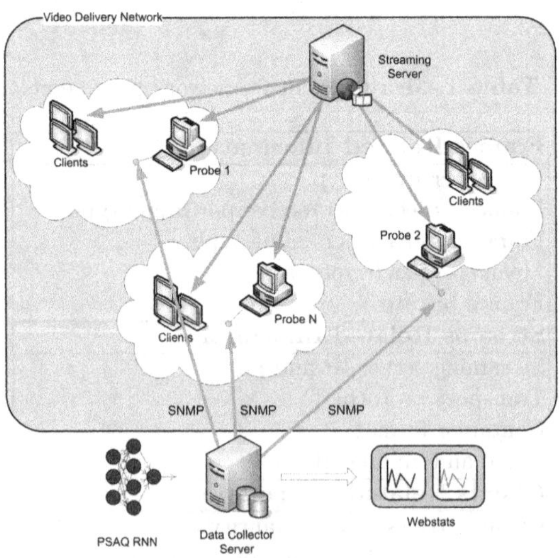

Fig. 1. Global architecture - The streaming server, the probes, the data collector server, the PSQA Learning Tool and the Webstat

streaming server. In our test scenario we chose the first option, by means of a modified VLC server. In a more realistic situation it may be not possible to use our own VLC servers; in that case, a preprocessed video (with user data) is needed to use with a generic stream server.

Probes (VLC players). Probes are VLC players modified in order to measure some specific information. They are strategically located inside the VDN. Basically, a probe is a VLC player with supplementary modules for coordination and data calculation, a SNMP module and a Logs module. They allow to capture and parse the user-data, to measure and to calculate generic information (like the start and stop of the stream), to offer realtime reports through SNMP and to manage a set of rotating logs files with all the relevant probe information.

Data Collector Server. The data collector server is in charge of gathering the probes's information. This application polls each one of the probes in the VDN (with some periodicity) and saves the data on a Round Robin Database (RRD). In this case, the data collector server polls the probes every 10 seconds and one RRD is created per active stream.

PSQA Tool. In this subsection we study how the frame loss process affects the perceptual quality (as measured by the PSQA technique). The first step was to apply the PSQA technique, as explained in Section 2. For this, we chose four MPEG-2 video sequences, of about 10 seconds each, with sizes between 1.5 MB and 2.8 MB. For each sequence, we generated twenty five different evaluation points, where each evaluation point is defined by a loss rate value chosen at random with an uniform distribution between 0.0 and 0.2, and a mean loss burst size value chosen at random with an uniform distribution between 0.0 and 10.0 (the actual process is a little bit more complex but this does not change the essential aspects of the method, see [8] for more details). For each of the evaluation points, we used a simple Markov chain (a simplified Gilbert model [8, 9]) to simulate a frame drop history which was applied to the original video sequences. In this way, we obtained one hundred modified video sequences with variable quality levels. Then, a group of five experts evaluated the sequences and the MOS for each of the copies was computed, following the ITU-R BT.500-11 norm [10] (see Figure 2(a)). These MOS were scaled into a quality metric in the range $[0, 1]$. Finally, we employed the MOS value for each of the design points as inputs in order to calibrate a Random Neural Network (RNN). After trained and validated, the RNN provides a function of two variables, *LR* and *MLBS*, mapping them into perceived quality (on a $[0, 1]$ range). In Figure 2(b) we can see the obtained function. For ease of reading, we extrapolated the curve to the borders, but observe that the data are accurate and used on an internal region ([1%, 15%] for *LR*, and [1, 4] for the *MLBS*). In particular, we can observe that quality is monotone in the two variables, and particularly increasing with the *MLBS*, meaning that, in this losses range, humans prefer sequences where losses are concentrated over those where losses are spread through the flow.

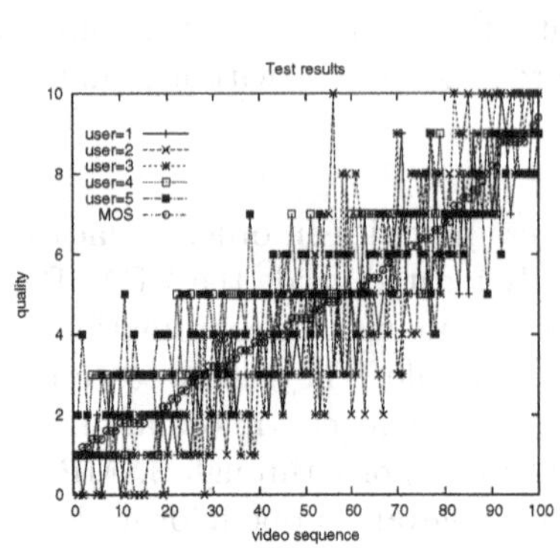

(a) Subjective test results for building the PSQA metric

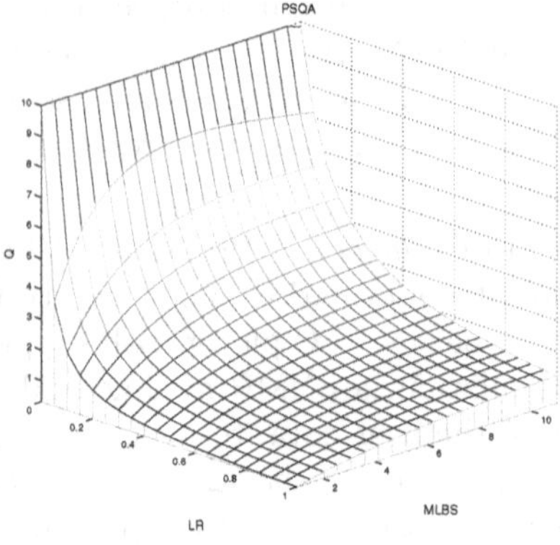

(b) The PSQA curve in our setting

Fig. 2. The subjective test input of our PSQA technique, and the PSQA function (mapping *LR* and *MLBS* into perceived quality) after trained and validated

Observe that training is only made once, when building the tool. In operation, the use of PSQA is very simple: the probes sends statistical information about the frame loss rate *LR* and the mean loss burst size *MLBS* in a short period to the data collector, to be evaluated by the PSQA Module. The period size can be arbitrarily defined for the specific application, and it is usually recommended to make it quite short, in order to use PSQA as an *instantaneous* quality value.

Webstat. `Webstat` is an application designed to present the data gathered by the data collector server to administrators and managers of the VDN. It offers reports at different levels and types of views. It is possible to generate reports focused on a particular stream or on a specific client (where there could be more than one active stream), or perhaps on the entire network (where there could be more than one client). Finally, it is possible to display the results at the frame level, possibly per frame type (I, P, B), or at the PSQA level.

Implementation. As mentioned before, both the probes and the streaming server are based on a VLC Player. The `libavcodec (ffmpeg)` library was used in order to work with MPEG-2 and MPEG-4 video specifications. As container format we worked with MPEG-TS using the functionalities provided by `libdvbps`. We streamed over HTTP and UDP using VLC internal modules. In order to report the probes's measures through SNMP we used the Net-SNMP library. We used the RRDtool to generate the statistical graphs. The data collector server was written in PHP and we used the following PHP extensions: `php4-snmp` and `php4-rrdtool`. `Webstat` is also written in PHP; it uses MySQL as relational database and it runs over Apache. All libraries and applications mentioned above are free-software and they can be executed in Microsoft Windows, Unix, Linux, etc.

4 Evaluation and First Results

To illustrate the platform possibilities, we tested it in some simple scenarios. We show here some examples of the obtained results. The scenarios concern the basic case of a VDN over the Internet.

Testing Scenarios. We simulated a VDN and we used our tool to monitor it. We used three streaming servers located at the same network, ten clients located at another network, one computer carrying out the routing function between both networks and a second computer where the data collector server and the `webstat` application are run.

We considered three different scenarios: one of them without losses (NoLS), another one with server failures (SLS) and a last one with congestion (packet losses) in the network (NLS). The first scenario consisted of a streaming server who sends traffic to three clients over a network without losses. In the second scenario, some video frames are eliminated from the stream by using a specific VLC server. This scenario was composed of one streaming server and two clients. Finally, in the third configuration we eliminated IP packets in the routing computer (congestion losses simulation). The losses were generated in an uniform way using the `netem` Linux module. This scenario was composed of one streaming server and five clients. In Table 2 we present information about each of the streams used in the evaluations: the transport protocol, the video specification and the video bitrate of the streaming, the scenario where the test case was executed and, if it matters, the loss rate. When HTTP is the transport protocol, a maximum bandwidth is set in the network. This is in order to make the IP packet losses have a direct impact on the quality of the stream; otherwise the lost packets would be retransmitted and will only be affecting the bandwidth consumed in the network. In the case of sequences coded at 512 Kbps, a bandwidth of 717 Kbps is set in the network; for sequences coded at 1024 Kbps, the maximum bandwidth is 1280 Kbps.

Table 2. Scenarios Configuration. Each test case correspond to a client receiving a stream. Each scenario correspond to a failure situation: without losses (NoLS), with server failures (SLS) and with packet losses in the network (NLS).

Test Case	Protocol	Video Specification	Video Bitrate (Kbps)	Scenario	Loss Rate
1	UDP	MPEG-2	1024	NoLS	-
2	HTTP	MPEG-2	1024	NoLS	-
3	HTTP	MPEG-4	512	NoLS	-
4	HTTP	MPEG-4	512	SLS	0.50
5	HTTP	MPEG-4	1024	SLS	0.30
6	UDP	MPEG-2	512	NLS	0.30
7	UDP	MPEG-2	1024	NLS	0.30
8	UDP	MPEG-4	1024	NLS	0.30
9	HTTP	MPEG-2	1024	NLS	0.02
10	HTTP	MPEG-4	1024	NLS	0.04

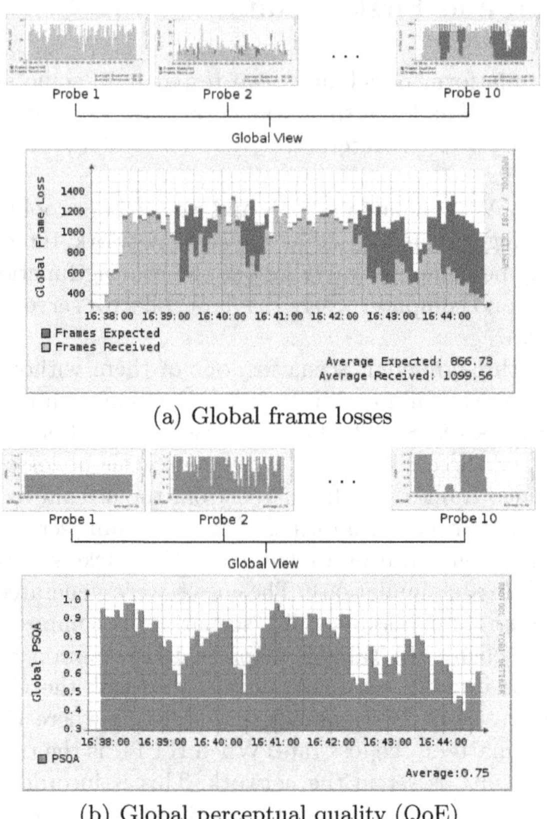

(a) Global frame losses

(b) Global perceptual quality (QoE)

Fig. 3. Typical global view of the entire Video Delivery Network

Obtained Results. As mentioned in Section 3 and as shown in Fig. 3, the developed application lets us to analyze the results at different levels and providing different views. In Fig. 3(a) we show a global view of the relative error measured during the evaluation at the frame level. This information can also be provided on a per-frame type basis (I-Frame, P-Frame, B-Frame).

In Fig. 3(b) we show the evolution of quality (PSQA) with time, normalized to numbers in the interval $[0, 1]$. In Table 3 we present some values obtained when executing the simulations. As it can be observed, there is no visible relationship between IP packet losses and frame losses. This is because this relationship depends on various factors: the video specification, the video bitrate, and the specific player software that processes the stream. However, there is a clear relationship between the loss rate and the PSQA value: the higher the loss rate the lower the quality perceived by the users. Last, Fig. 4(a) shows the frame losses measured during a test case where server failures occur and Fig. 4(b) shows the frame losses measured during a test case with network losses.

Table 3. Obtained Results

Test Case	Protocol	Specified Loss Rate	Measured Frame Loss Rate	Mean PSQA
1	UDP	-	-	1.00
2	HTTP	-	-	1.00
3	HTTP	-	-	1.00
4	HTTP	0.50 (Frame level)	0.47	0.42
5	HTTP	0.30 (Frame level)	0.29	0.61
6	UDP	0.30 (IP level)	0.09	0.79
7	UDP	0.30 (IP level)	0.19	0.66
8	UDP	0.30 (IP level)	0.33	0.38
9	HTTP	0.02 (IP level)	0.08	0.79
10	HTTP	0.04 (IP level)	0.07	0.89

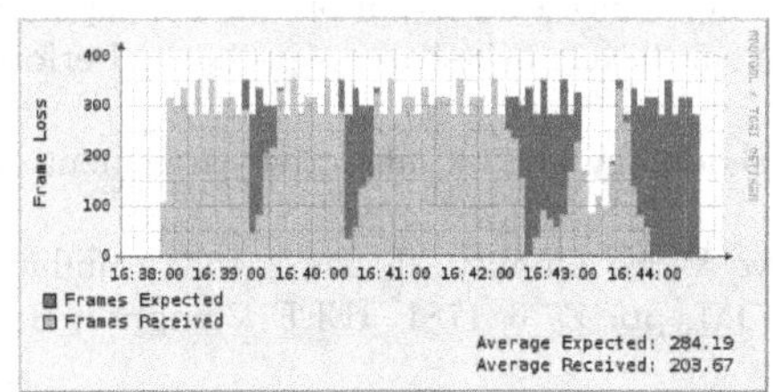

(a) Server failure example (5th test case)

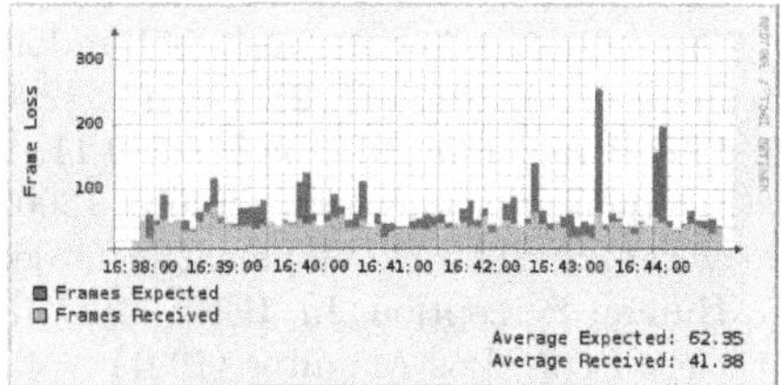

(b) Network loss example (8th test case)

Fig. 4. Typical stream view of the Video Delivery Network

5 Conclusion

This paper presents an effective monitoring and measuring tool that can be used by VDN managers and administrators to assess the streaming quality inside the network. With this tool it is possible to automatically monitor different sets of parameters of the streams, in real-time if necessary, including the perceived quality as seen by the final users, thanks to our improved version of the recently proposed PSQA technology. PSQA provides an accurate approximation of the QoE (Quality of Experience) and, to the best of our knowledge, our tool is the only one that is able to evaluate perceived quality continuously at arbitrarily chosen points in the network.

Another important feature of our tool is that it is not dependent on the considered VDN. It was designed as a generic implementation, in order to be able to use it with multiples VDN architectures. Moreover, it can be associated with most common management systems since it is built over the SNMP standard. Another feature is that the induced overhead is negligible. Finally, the tool is a free-software application that can be executed on several operating systems.

References

1. DSL Forum Technical Work WT-126: Video services quality of experience (qoe) requirements and mechansims (2007)
2. Evertz Microsystems Ltd. - Manufacturer of High Quality Film and Digital Broadcast Equipment (2007), `http://www.evertz.com/`
3. Tandberg Home page (2007), `http://www.tandberg.com/`
4. Cisco Systems, Inc. (2007), `http://www.cisco.com/`
5. Alcatel-Lucent Home (2007), `http://www.alcatel-lucent.com/`
6. Siemens AG (2007), `http://www.siemens.com/`
7. IETF Network Working Group: An Architecture for Describing Simple Network Management Protocol (SNMP) Management Frameworks (RFC 3411) (2002)
8. Mohamed, S., Rubino, G.: A Study of Real–time Packet Video Quality Using Random Neural Networks. IEEE Transactions On Circuits and Systems for Video Technology 12(12), 1071–1083 (2002)
9. Mohamed, S., Rubino, G., Varela, M.: Performance evaluation of real-time speech through a packet network: a Random Neural Networks-based approach. Performance Evaluation 57(2), 141–162 (2004)
10. ITU-R Recommendation BT.500-11: Methodology for the subjective assessment of the quality of television pictures (2002)
11. Voran, S.: The Development of Objective Video Quality Measures that Emulate Human Perception. In: IEEE GLOBECOM, pp. 1776–1781. IEEE Computer Society Press, Los Alamitos (1991)
12. van den Branden Lambrecht, C.: Perceptual Models and Architectures for Video Coding Applications. PhD thesis, EPFL, Lausanne, Swiss (1996)
13. Gelenbe, E.: Random Neural Networks with Negative and Positive Signals and Product Form Solution. Neural Computation 1, 502–511 (1989)
14. VideoLan home page (2007), `http://www.videolan.org`
15. IETF Network Working Group: Management Information Base (MIB) for the Simple Network Management Protocol (SNMP) (RFC) 3418) (2002)
16. IETF Network Working Group: Structure of Management Information Version 2 (RFC 2578) (1999)
17. IETF Network Working Group: Textual Conventions for SMIv2 (RFC 2579) (1999)
18. IETF Network Working Group: Conformance Statements for SMIv2 (RFC 2580) (1999)

Measurement and Analysis of Intraflow Performance Characteristics of Wireless Traffic

Dimitrios P. Pezaros, Manolis Sifalakis, and David Hutchison

Computing Department, Infolab21, Lancaster University, Lancaster, UK, LA1 4WA
{dp,mjs,dh}@comp.lancs.ac.uk

Abstract. It is by now widely accepted that the arrival process of aggregate network traffic exhibits self-similar characteristics which result in the preservation of traffic burstiness (high variability) over a wide range of timescales. This behaviour has been structurally linked to the presence of heavy-tailed, infinite variance phenomena at the level of individual network connections, file sizes, transfer durations, and packet inter-arrival times. In this paper, we have examined the presence of fractal and heavy-tailed behaviour in a number of performance aspects of individual IPv6 microflows as routed over wireless local and wide area network topologies. Our analysis sheds light on several questions regarding flow-level traffic behaviour: whether burstiness preservation is mainly observed at traffic aggregates or is it also evident at individual microflows; whether it is influenced by the end-to-end transport control mechanisms as well as by the network-level traffic multiplexing; whether high variability is independent from diverse link-level technologies, and whether burstiness is preserved in end-to-end performance metrics such as packet delay as well as in the traffic arrival process. Our findings suggest that traffic and packet delay exhibit closely-related Long-Range Dependence (LRD) at the level of individual microflows, with marginal to moderate intensity. Bulk TCP data and UDP flows produce higher Hurst exponent estimates than the acknowledgment flows that consist of minimum-sized packets. Wireless access technologies seem to also influence LRD intensity. At the same time, the distributions of intraflow packet inter-arrival times do not exhibit infinite variance characteristics.

Keywords: LRD, Hurst exponent, heavy-tailed distribution, ACF.

1 Introduction

Seminal measurement studies during the last fifteen years have demonstrated that data communications networks' traffic is self-similar (statistically fractal) in nature remaining bursty over a wide range of timescales. These findings advocated that statistical properties of the (aggregate) network traffic, when viewed as time series data, remain similar irrespective of the time scale of observation, and were in sharp contrast with the up till then commonly employed Poisson and Markovian models which were based on exponential assumptions about the traffic arrival process. A characteristic of self-similar processes is that they often exhibit long memory, or

D. Medhi et al. (Eds.): IPOM 2007, LNCS 4786, pp. 143–155, 2007.

Long-Range Dependence (LRD), signifying that their current state has significant influence on their subsequent states far into the future. Consequently, values at a particular time are related not just to immediately preceding values, but also to fluctuations in the remote past. Hence, high variability in the behaviour of self-similar processes is preserved over multiple time scales. Pioneering work has focused on the measurement and characterisation of LAN [13], WAN [17], and transport/application-specific traffic [6], having the traffic arrival process at a single network (edge) point as the common primary quantity of interest. The revealed LRD properties of aggregate network traffic have been subsequently linked to heavy-tailed, infinite variance phenomena at the level of individual source-destination pairs, represented by ON/OFF sources and packet trains models whereby a source alternates between active (ON-period) and idle (OFF-period) states [22][23]. These are in turn attributed to distributed systems' objects and file sizes being heavy-tailed, a property that has been shown to be preserved by the protocol stack and mapped to approximate heavy-tailed busy periods at the network layer [15][16]. However, the existence (or otherwise) of self-similar and/or heavy-tailed behaviour within performance facets of individual microflows has not received much attention to date. This is partly due to the majority of individual connections in the Internet being short-lived, and therefore their behaviour over multiple time scales can be studied only through aggregation. At the same time, the vast majority of bytes over the Internet are carried within a relatively small number of very large flows which are sufficiently long-lived for the temporal evolution of their intraflow properties to be investigated [4]. It is henceforth feasible to examine whether the self-similar characteristics of aggregate traffic resulted from the superimposition of numerous ON/OFF sources are also manifested at the level of long-lived individual traffic flows. Characterising the long-term flow behaviour and revealing possible burstiness preservation properties therein can prove instrumental for network resource management and accountability purposes, since end-to-end flows are the primary entity subjected to open and closed-loop network control. Similar to proposals advocating small buffer capacity/large bandwidth resource provisioning when input traffic is self-similar, relevant characterisation of end-to-end flow behaviour and performance aspects therein can form the basis for designing adaptive flow control algorithms to operate at multiple timescales and take into consideration potential correlation between distant events/values throughout the lifetime of a flow. From a measurement point of view, the comparative analysis of temporal flow behaviour is also important since it can reveal certain idiosyncrasies potentially linked to the operation of the different transport control algorithms. Differences in the long-term behaviour of diverse traffic flows can hint to additional causality relationships between burstiness intensity and flow control, as well as the levels of traffic multiplexing in the network.

Likewise, packet delay is one of the most commonly employed metrics to assess the service quality levels experienced by an end-to-end flow. In contrast to traffic arrivals, delay indicates how traffic is routed between two network nodes and is among the primary performance aspects whose absolute value and temporal variations (in either the unidirectional or the round-trip representation) are attempted to be controlled by transport mechanisms and kept within certain ranges depending on individual applications' requirements. Links between network traffic self-similarity and the temporal intraflow delay behaviour can identify the degree of penetration of high variability to different facets of network performance, and the relative level of causality between performance, end-to-end flow control, and traffic multiplexing in the network. A few studies on fractal analysis of packet delay have reported non-stationary LRD in round-trip delay of synthetic UDP traffic [3], and in aggregate NTP/UDP flows [14], yet the burstiness preservation relationships between the different transport mechanisms and the unidirectional contributors of the end-to-end delay of individual flows have not been investigated.

In this paper, we have quantified the burstiness preservation properties of a set of end-to-end unidirectional IPv6 traffic flows routed over IEEE 802.11 and GPRS service networks, two media which themselves exhibit highly variable performance characteristics [10][5]. We have comparatively examined the presence of Long-Range Dependence (LRD) in the intraflow traffic arrival process and in per-packet unidirectional delay, and we have investigated the extent to which transport control, packetisation, and wireless access mechanisms influence its intensity. In addition, we have analysed the tail behaviour of packet inter-arrival times at the individual flow level, and revealed the absence of infinite variance phenomena therein. In section 0 we provide the definition and mathematical formulation of self-similarity and LRD, their interrelation, and a brief discussion on the estimators used to identify and quantify LRD on empirical time series. Section 0 includes an outline our measurement methodology and a description of the wireless experimental infrastructure over which measurements were conducted. In section 0 we present and discuss in detail the results of measurement analysis, and we provide possible explanations and interpretations of our findings. We comment on the similarity between LRD in the per-flow traffic and unidirectional delay, and on the different levels of LRD exhibited by diverse application flows and over the different wireless topologies. We also compare and contrast the tail behaviour of per-flow packet inter-arrival times to heavy-tailed distributions. Section 0 concludes the paper.

2 Self-similarity and Long-Range Dependence: Definitions and Estimation

A stochastic process or time series *Y(t)* in continuous time $t \in \mathbb{R}$ is self-similar with self-similarity (Hurst) parameter $0 < H < 1$, if for all $\alpha > 0$ and $t \geq 0$,

$$Y(t) =_d a^{-H} Y(\alpha t).$$

Self-similarity describes the phenomenon of a time series and its time-scaled version following the same distribution after normalizing by α^{-H}. It is relatively straightforward to show [2] that this implies that the autocorrelation function (ACF) of the stationary increment process $X(t) = Y(t) - Y(t-1)$ at *lag k* is given by

$$\rho(k) = \frac{1}{2}((k+1)^{2H} - 2k^{2H} + (k-1)^{2H}),\ k \geq 1.$$

In addition, for $0 < H < 1$, $H \neq \frac{1}{2}$, it can be shown that $\lim_{k \to \infty} \rho(k) = H(2H-1)k^{2H-2}$, and in particular for the case $0.5 < H < 1$, *ρ(k)* asymptotically behaves as $ck^{-\beta}$ for

$0 < \beta < 1$, where $c > 0$ is a constant, $\beta = 2 - 2H$ [2][16]. This implies that the correlation structure of the time series is asymptotically preserved irrespective of time aggregation, and the autocorrelation function decays hyperbolically which is the essential property that constitutes it not summable:

$$\sum_{k=1}^{\infty} \rho(k) = \infty.$$

When such condition holds, the corresponding stationary process *X(t)* is said to be *Long-Range Dependent (LRD)*. Intuitively, this property implies that the process has infinite memory for $0.5 < H < 1$, meaning that the individually small high-lag correlations have an important cumulative effect. This is in contrast to conventional short-range dependent processes which are characterised by an exponential decay of the autocorrelations resulting in a summable autocorrelation function. LRD causes high variability to be preserved over multiple time scales and is one of the manifestations of self-similar processes alongside non-summable spectral density for frequencies close to the origin and slowly decaying variances. This latter characteristic of self-similar and LRD processes can be disastrous for classical tests and prediction of confidence intervals. The variance of the arithmetic mean decreases more slowly than the reciprocal of the sample size, behaving like $n^{-\beta}$ for $0 < \beta < 1$, instead of like n^{-1} which is the case for processes whose aggregated series converge to second-order pure noise [2][13]. Therefore, usual standard errors derived for conventional models are wrong by a factor that tends to infinity as the sample size increases. Two theoretical models that have been used to simulate LRD is the fractional Gaussian noise (fGn) which is the stationary increment process of fractional Brownian motion (fBm), and fractional ARIMA processes that can simultaneously model the short and long term behaviour of a time series.

A number of estimators [21][13] have been extensively used in multidisciplinary literature for LRD detection and quantification by estimating the value of the Hurst exponent; as $H \to 1$ the dependence is stronger. They are classified in time-domain and frequency-domain estimators, depending on the methodology they employ to estimate *H*. Time-domain estimators are based on heuristics to investigate the evolution of a statistical property of the time series at different time-aggregation levels. They are hence mainly used for LRD detection, rather than the exact quantification of the phenomenon. Frequency-domain estimators focus on the behaviour of the power spectral density of the time series. In this paper, we have employed two time-domain estimators (the *aggregated variance* and the *rescaled adjusted range* methods) to detect whether our measured quantities exhibit LRD characteristics ($H > 0.5$), and we have subsequently focused on the more robust frequency-domain *Whittle* estimator for the exact LRD quantification. Whittle is a maximum likelihood type estimate which is applied to the periodogram of the time series and it provides an asymptotically exact estimate of *H* and a confidence interval. However, it presupposes that the empirical series is consistent with a specific process (e.g. fGn) whose underlying form must be provided, hence its use with time series that have already been shown to be LRD (by other means) is strongly advisable. Wavelet-based LRD estimation [1] has also been developed whose accuracy is comparable to Whittle, yet it has been lately shown to consistently overestimate the Hurst exponent on synthesized LRD series and hence Whittle was preferred for this study [12].

3 Measurement Methodology and Experimental Environment

In-line measurement [18] has been employed to instrument a representative set of IPv6 traffic flows as these were routed over diverse wireless topologies during one week, in November 2005 [19]. The technique exploits extensibility features of IPv6, to piggyback measurement data in Type-Length-Value (TLV) structures which are then encapsulated within an IPv6 *destination options* extension header and carried between a source and a destination. Being encoded as a native part of the network-layer header, inline measurement is potentially applicable to all traffic types carried over the IPv6 Internet infrastructure. Destination options extension headers in particular, are created at the source and are only processed at the (ultimate) destination node identified in the destination address field of the main IPv6 header. Hence, their presence does not negatively impact the option-bearing datagrams at the intermediate forwarding nodes [18]. For the purposes of this study, 32-bit Linux kernel timestamps were encoded in an appropriate TLV structure to record time T_{DEP} immediately before a packet is serialised at the NIC of the source IPv6 node, and time T_{ARR} as soon as the packet arrives in the destination IPv6 node's OS kernel. The intraflow traffic arrival process has been calculated as the number of packets (or bytes) arriving at the destination within disjoint subintervals throughout the flow duration. The end-to-end unidirectional delay for a given packet P is calculated as $D = T^P{}_{ARR} - T^P{}_{DEP}$. Moreover, the inter-arrival time between two successive packets P_i and P_{i+1} is computed as $IA = T_{ARR}^{P_{i+1}} - T_{ARR}^{P_i}$. For the purposes of the delay measurement the two end-systems synchronised using the Network Time Protocol (NTP) with a common stratum 1 server through additional high-speed wired network interfaces, in order to avoid NTP messages competing with the instrumented traffic over the bottleneck wireless links. The NTP daemon was allowed sufficient time to synchronise prior to the experiments until it reached a large polling interval. The offset reported by NTP was always at least one order of magnitude smaller with respect to the minimum one-way delay observed. All the delay traces were empirically examined against negative values as well as against linear alterations (trend) of the minimum delay over time. None of these offset/skew-related phenomena were experienced.

Instrumented traffic consisted of moderate-size bulk TCP transfers and CBR UDP video streaming flows. Measurements have been conducted end-to-end over two diverse wireless service networks over the Mobile IPv6 Systems Research Laboratory (MSRL) infrastructure. MSRL includes a wireless cellular network as well as a combination of 802.11 technologies and it comprises a real service infrastructure, as shown in **Fig. 1**. The measurements were carried out between a host machine connected to MSRL's wired backbone network (Linux 2.4.18; Intel 100BaseT adapter) and a host machine with multiple wireless interfaces (Linux 2.4.19; NOKIA D211 combo PCMCIA 802.11b/GPRS adapter), connected through the 802.11b/g campus-wide network and through the GPRS/GSM W-WAN network. The W-LAN infrastructure is part of Lancaster University campus wireless network, and includes 802.11b and 802.11g. Although the nominal speed for 802.11b is 11Mb/s, it has been observed that due to interference with other appliances operating at the same frequency band (2.4 GHz), the cards often fallback to 5.5, 2, and 1 Mb/s.

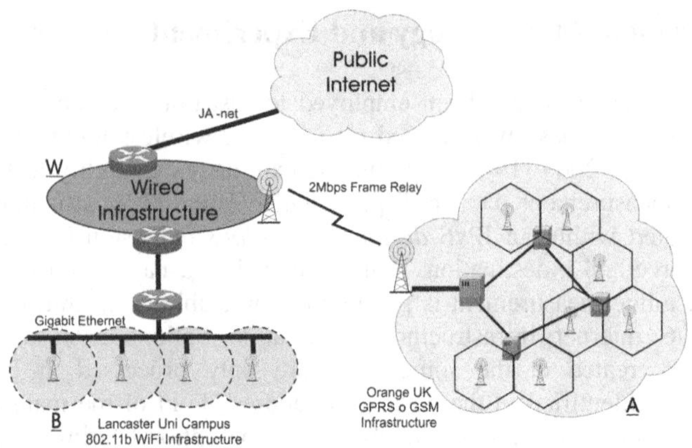

Fig. 1. Wireless experimental environment (MSRL infrastructure)

The W-WAN network is the Orange UK GPRS/GSM service network, practically allowing for speeds of up to 20/50Kb/s (up/downlink), due to asymmetric slot allocation. Connectivity between Orange UK and the MSRL backbone is served by a 2Mb/s wireless Frame Relay point-to-point link.

4 Measurement Analysis and Results

Packet-level measurements were taken upon arrival of each packet to its destination, hence at irregular time instants. In order to covert the traces to time series data, they were discretised into equally-sized bins based on normalised packet arrival time. Unidirectional delays and inter-arrival times of multiple packets arriving within each bin were averaged, and the mean values were considered for the particular bin. Although this process inevitably smooths out short-term variations, bin size was carefully selected for each flow to contain as few packets as possible, while at the same time avoiding empty bins. The time series' lengths varied from 2^9 to 2^{12} which has been reported sufficient for Hurst exponent estimates with less than 0.05 bias and standard deviation [7]. A number of sanity tests and calibration measures have been employed to tune the LRD estimation process by eliminating time series effects such as periodicity, trend and short-range correlations which are known to constitute LRD estimation error prone. Trends and non-stationarities have been checked against during pre-processing, while periodicity and short-term correlations have been neutralised using the randomised buckets method to perform internal randomisation to the signal [9]. Time-domain (detection) estimators which operate on the aggregated time series can suffer from the limited number of samples available at high aggregation levels. We have employed oversampling with digital low-pass filtering to increase the sampling rate of the time series and hence refine their estimates. For further details regarding techniques for tuning the LRD estimation process and their effects, the interested reader is referred to [12][20].

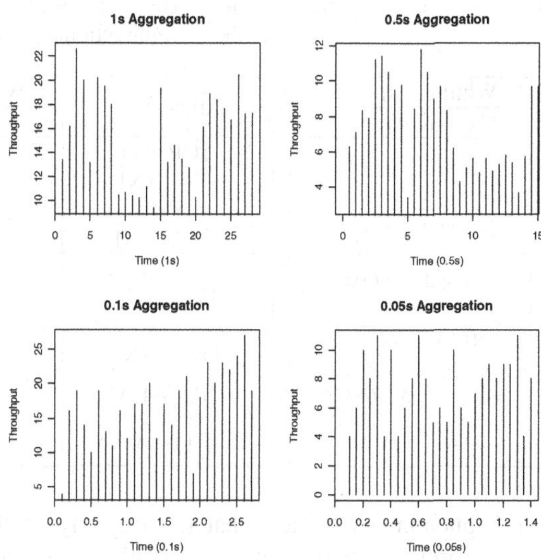

Fig. 2. Stochastic Self-Similarity – burstiness preservation across time scales of 50 ms, 100 ms, 500 ms, and 1000 ms for the bulk TCP data path over the W-LAN network

4.1 LRD Behaviour in Traffic Arrivals and Unidirectional End-to-End Delay

The two intraflow properties analysed for LRD behaviour were the traffic arrival process and the unidirectional end-to-end packet delay. These were comparatively examined for TCP data, TCP reverse, and CBR UDP flows routed over the two diverse wireless topologies. We have used the LRD detection (time-domain) estimators on all time series data which all reported that traffic arrivals and one-way delay exhibit LRD at various intensities, and produced Hurst exponent estimates $0.5 < H < 1$. We subsequently applied the Whittle estimator on the data to quantify the levels of LRD with high confidence. **Fig. 2** indicatively shows the burstiness preservation of the data path of a bulk TCP flow over the W-LAN topology at varying time scales, demonstrating the absence of a characteristic size of a burst. Traffic throughput is shown in packets which are all MSS-sized and hence byte-throughput is linearly identical. The upper left plot shows a complete presentation of the time series using one-second bins. Then, the bottom left plot shows the same time series whose first 3-second interval is "blown up" by a factor of ten, and the truncated time series has time granularity of 100 ms. Likewise, the rightmost plots show parts of the time series with an equivalent number of samples for time granularities of 500 and 50 ms, respectively. Overall, the plots show significant bursts of traffic at different levels spanning almost three orders of magnitude. **Table 1** and **Table 2** show the Whittle estimate (and 95% confidence interval) of the Hurst exponent for traffic arrivals and unidirectional delay, respectively, for all the different flows routed over the W-LAN and W-WAN networks. Their comparative examination yields some very interesting observations.

Table 1. Hurst exponent estimates – Whittle method: Traffic arrivals

Microflow	Whittle Estimator
	H est. & 95% C.I.
TCP data [W-LAN]	0.714 ± 0.005
TCP data [W-WAN]	0.554 ± 3.47e-05
TCP reverse [W-LAN]	0.584 ± 0.004
TCP reverse [W-WAN]	0.534 ± 0.001
UDP [W-LAN]	0.534 ± 3.37e-05
UDP [W-WAN]	0.697 ± 0.001

Table 2. Hurst exponent estimates – Whittle method: Unidirectional packet delay

Microflow	Whittle Estimator
	H est. & 95% C.I.
TCP data [W-LAN]	0.739 ± 0.025
TCP data [W-WAN]	0.599 ± 0.15
TCP reverse [W-LAN]	0.552 ± 0.007
TCP reverse [W-WAN]	0.528 ± 0.014
UDP [W-LAN]	0.687 ± 0.003
UDP [W-WAN]	0.742 ± 0.20

It is evident in both phenomena (tables) that the majority of the unidirectional flows over the two media independently, show marginal to moderate LRD intensity with Hurst exponent values for some of them close to those of short-range dependent processes, differing by less than 0.1. This fact reinforces the argument of LRD being intensified by the aggregation of traffic. However, there are cases of individual flows which suggest dependency between LRD, traffic type and wireless medium. For bulk TCP data over W-LAN, both traffic arrival (whose burstiness preservation was shown in **Fig. 2**) and end-to-end delay exhibit considerable LRD manifested by Hurst values greater than 0.71. This is in contrast to the same type of traffic routed over W-WAN for which both properties assume marginal intensity values less than 0.6. The opposite behaviour with respect to the two wireless media is observed for constant bit rate UDP traffic. Over W-LAN, UDP traffic does not assume considerable LRD, whereas moderate intensity is suggested for UPD over W-WAN with an estimated Hurst exponent close to 0.7. The same relationship holds for the packet delay behaviour of the UDP flows over the two media, although the absolute Hurst estimates are in both cases larger than those of the traffic process. The acknowledgment path of bulk TCP connections over both media is characterised by smaller intensity than the corresponding data path, and overall marginal LRD.

When comparing the Hurst estimates of the per-flow traffic behaviour with those of the corresponding unidirectional delay, it is apparent that there is a considerably close agreement between their LRD intensities. This implies that although traffic arrival process and unidirectional delay are metrics describing different aspects of network dynamics (the former describes how traffic is delivered at a single network node while the latter describes how traffic is routed between two nodes), they are both influenced by common sets of parameters. Hurst estimates of the two processes for each flow/medium combination lie within a range which is smaller than the wider 95% confidence interval of the two. The only exception is the TCP acknowledgment path over the W-LAN topology, where the Hurst estimates for traffic and packet delay differ by 0.032 while the wider 95% C.I. of the two (delay) is 0.007. Overall, Whittle performs better on the traffic arrival process by producing narrow 95% C.I.s for the

Hurst estimates on the order of 10^{-3} or better. For unidirectional packet delay the corresponding C.I.s are on the order of 10^{-2} or better, however in two cases, the width of the 95% C.I.s can put even the existence of marginal LRD under doubt.

It is well accepted that traffic self-similarity arises through the aggregation of multiple traffic sources since its causality relationship with the heavy-tailed property of the ON/OFF sources model is based on limit theorems for large number of sources and large temporal intervals [22][23]. However, the per-flow indications of LRD for traffic as well as for the packet delay raise some interesting issues. The two wireless configurations over which the measurement was conducted can be safely assumed to operate as access networks where clients are mainly downloading content and not uploading, hence the download path is more congested than the reverse direction. At the same time, the W-LAN network was lightly utilised during the time of the experiments, as this was indicated by APs' client association logs. Therefore, the stronger LRD intensity exhibited by the TCP data flow over W-LAN yields some dependency between LRD and the packetisation/congestion control algorithm operating on the flow, irrespective of traffic aggregation over the medium. The fact that similar LRD levels are not seen for the TCP data path over W-WAN hints towards relationship between traffic behaviour and link-local delivery mechanisms. The dense protocol stack of GPRSoGSM which to some extent replicates TCP's reliable transmission seems to neutralise the effect that transport-layer congestion control has on the long-range behaviour of traffic. The higher Hurst estimates produced for both TCP data and UDP flows, as opposed to the (lower) estimates of the TCP reverse flows over both wireless media signifies dependency between LRD behaviour and packet size. It is worth noting that TCP data and UDP flows consisted of constant-size packets of 1440 and 544 bytes, respectively, while TCP reverse flows consisted of (mainly) 56 and (fewer) 84-byte packets. All sizes exclude network and link-layer headers. Apart from the Hurst exponent estimates, accompanying indisputable evidence regarding long-memory behaviour of per-flow data are computed sample statistics such as ACF which exhibit nontrivial correlations at large lags. **Fig. 3** indicatively shows the correlation structures of traffic arrivals from a selection of flows with different LRD intensity whose estimates appear in **Table 1**. The ACFs demonstrate the different correlation structures between time series with considerably different Hurst estimates. The plots also show the effect of the randomised buckets with internal randomisation which were employed to neutralise short-range correlations and periodicities. It is evident that ACFs follow asymptotic decay at various levels, which is unsurprisingly more intense for flows with larger Hurst estimates. It is interesting to note that even for the TCP reverse flow whose estimated Hurst value is considerably small (<0.6), correlations seem non-degenerate since they mostly remain above the 95% C.I. of zero.

4.2 Tail Behaviour of Intraflow Packet Inter-Arrival Times

The heavy-tailed property of the transmission (or the idle) times of individual sources has been characterised as the main ingredient needed to obtain LRD characteristics ($H>0.5$) in aggregate traffic [22][23]. Heavy-tailness has since then been reported in a number of facets of traffic behaviour, including (web) file sizes, transfer times, burst lengths and inter-arrival times [6][8][17].

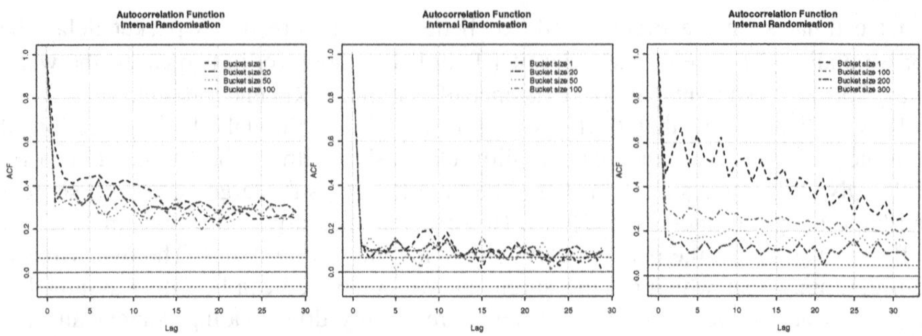

Fig. 3. Autocorrelation function (ACF) of TCP data [*H=0.714*] and TCP reverse [*H=0.584*] traffic over W-LAN, and UDP traffic [*H=0.697*] over W-WAN

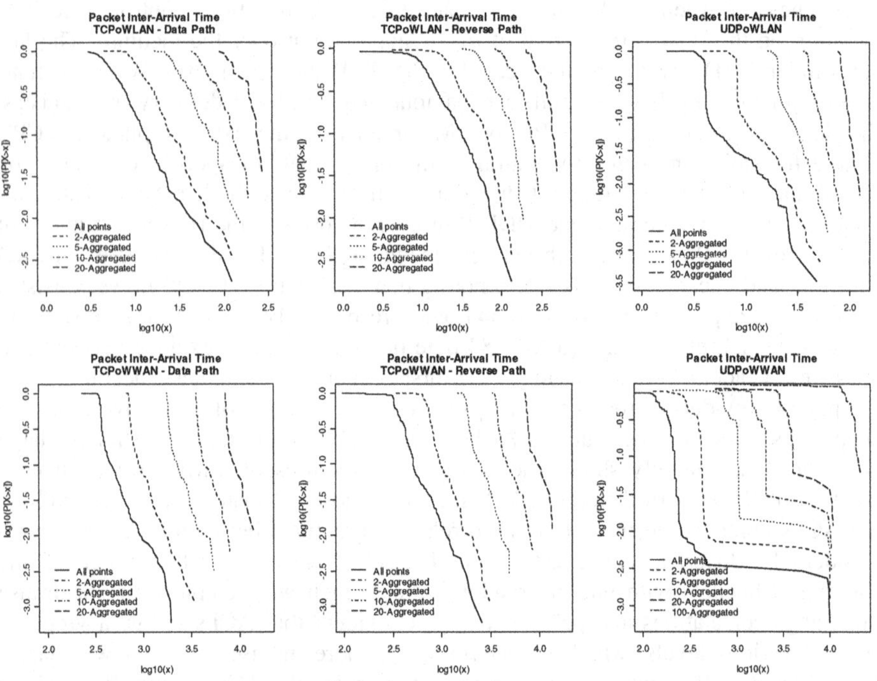

Fig. 4. LLCDs of intraflow packet inter-arrival times for TCP data, TCP reverse, and UDP traffic over W-LAN and W-WAN configurations

A distribution is heavy-tailed if $P[X > x] \sim x^{-\alpha}$, as $x \to \infty$, $0 < \alpha < 2$. That is, regardless of its behaviour for small values of the random variable, its asymptotic shape is hyperbolic. For $1 \leq \alpha < 2$, the distribution has infinite variance, whereas for $0 < \alpha < 1$, it has infinite mean as well. The tail index α can be empirically estimated using the log-log complimentary distribution (LLCD) plot and calculating its slope through least-squares regression for large values of the random variable above which the plot appears to be linear. We have used a test (also used in [6]) based on the Central Limit Theorem (CLT) to examine whether the packet inter-arrival times within our measured traffic flows exhibit infinite variance, and hence heavy-tailed characteristics. According to CLT, the sum of a large number of i.i.d. samples from any distribution with finite variance will tend to be normally distributed. Hence, for a distribution with finite variance, the slope of the LLCD plot of the *m-aggregated* dataset will increasingly decline as m increases, reflecting the distribution's approximation of a normal distribution. On the contrary, if the distribution exhibits infinite variance the slope will remain roughly constant with increasing aggregation levels (m). **Fig. 4** shows the CLT test for various aggregation levels applied to the packet inter-arrival times of all flows over the two wireless topologies. It is evident that for increasing aggregation levels the slope of the LLCD plots of packet inter-arrival times decreases, arguing against the infinite variance characteristics of heavy-tailed distributions like e.g. Pareto. Usually, higher aggregation levels are used for the CLT test however these could not be achieved given the length of the instrumented flows. Yet, the slope decrease becomes apparent even for slightly increasing aggregation levels and absence of infinite variance can be safely assumed. Indeed, the least-squares regression we performed on the LLCD plots yielded values $\alpha > 2$, and their overall shape was better described by (light-tailed) log-normal distributions. This finding is in accordance to other studies suggesting that log-normal distributions give better fit to many duration and inter-arrival distributions observed over the Internet, than heavy-tailed Pareto distributions do [8]. It has also been shown that this observed light-tailness is not contradictory to LRD of aggregate traffic [11].

5 Conclusion

We have examined the temporal behaviour of traffic performance characteristics at the level of individual, sufficiently long-lived flows, routed over diverse wireless networks. We have provided evidence of similar LRD intensity between the intraflow traffic arrival process and the unidirectional packet delay, demonstrating that LRD behaviour of aggregate traffic penetrates other measurable quantities at finer granularities, albeit in lesser intensities than the ~0.8 levels reported for traffic aggregates [13][6]. However, even for relatively small LRD intensity (Hurst) values, the ACFs indicate non-obviously-degenerate correlations at large lags. Through the comparative examination of refined Hurst estimates of intraflow traffic properties we identified the possibility of other network and traffic idiosyncrasies, such as transport control mechanisms, wireless access technologies and packet sizes, influencing the intensity of LRD behaviour. At the same time, we have shown the absence of infinite variance phenomena at the distributions of intraflow packet inter-arrival times. Although this study focused on IPv6 flows, similar behaviour is expected for IPv4 traffic, since both protocols assume the same packetisation mechanisms at their higher layers. Whether in practice intermediate routers treat IPv4 and IPv6 traffic identically and how this influences their performance deserves further experimental investigation. In addition, comparative performance analysis between flows routed over wireless technologies and their wired counterparts is to be further pursued.

References

[1] Abry, P., Veitch, D.: Wavelet Analysis of Long-Range Dependence Traffic. In: IEEE Transactions on Information Theory, IEEE Computer Society Press, Los Alamitos (1998)

[2] Beran, J.: Statistics for Long-Memory Processes. In: Monographs on Statistics and Applied Probability, Chapman and Hall, New York (1994)

[3] Borella, M.S., Brewster, G.B.: Measurement and Analysis of Long-Range Dependent Behaviour of Internet Packet Delay. In: Proceedings, IEEE Infocom 1998, pp. 497–504 (April 1998)

[4] Brownlee, N., Claffy, K.C.: Understanding Internet Traffic Streams: Dragonflies and Tortoises. IEEE Communications Magazine 40(10), 110–117 (2002)

[5] Chakravorty, R., Cartwright, J., Pratt, I.: Practical experience with TCP over GPRS. In: IEEE GLOBECOM 2002, Taiwan (2002)

[6] Crovella, M.E., Bestavros, A.: Self-similarity in World Wide Web traffic: Evidence and possible causes. IEEE/ACM Trans. on Networking 5(6), 835–846 (1997)

[7] Delignieres, D., Ramdani, S., Lemoine, L., Torre, K., Fortes, M., Ninot, G.: Fractal analyses for 'short' time series: a re-assessment of classical methods. Journal of Mathematical Psychology 50 (2006)

[8] Downey, A.B.: Lognormal and Pareto distributions in the Internet. Computer Communications 28(7), 790–801 (2005)

[9] Erramilli, E., Narayan, O., Willinger, W.: Experimental queuing analysis with long-range dependent packet traffic. IEEE/ACM Trans. on Networking 4(2), 209–223 (1996)

[10] Fotiadis, G., Siris, V.: Improving TCP throughput in 802.11 WLANs with high delay variability. In: ISWCS 2005. 2nd IEEE Int'l Symposium on Wireless Communication Systems, Italy (2005)

[11] Hannig, J., Marron, J.S., Samorodnitsky, G., Smith, F.D.: Log-normal durations can give long range dependence. In: Mathematical Statistics and Applications: Festschrift for Constance van Eeden, IMS Lecture Notes, Monograph Series, Institute of Mathematical Statistics, pp. 333-344 (2001)

[12] Karagiannis, T., Molle, M., Faloutsos, M.: Understanding the limitations of estimation methods for long-range dependence, Technical Report, University of California, Riverside, TR UCR-CS-2006-10245 (2006)

[13] Leland, W.E., Taqqu, M., Willinger, W., Wilson, D.V.: On the Self-Similar Nature of Ethernet Traffic (extended version. IEEE/ACM Trans. on Networking 2(1), 1–15 (1994)

[14] Li, Q., Mills, D.L.: On the long-range dependence of packet round-trip delays in internet. In: Proceedings of IEEE ICC 1998, IEEE Computer Society Press, Los Alamitos (1998)

[15] Park, K., Kim, G., Crovella, M.: On the Relationship Between File Sizes, Transport Protocols, and Self-Similar Network Traffic. In: IEEE International Conference on Network Protocols ICNP 1996, Ohio, USA, October 29- November 1, 1996, IEEE, Los Alamitos (1996)

[16] Park, K., Willinger, W.(eds.): Self-Similar Network Traffic and Performance Evaluation. John Wiley & Sons, New York (2000)

[17] Paxson, V., Floyd, S.: Wide-area traffic: The failure of Poisson modelling. IEEE/ACM Trans. on Networking 3(3), 226–244 (1994)

[18] Pezaros, D.P., Hutchison, D., Garcia, F.J., Gardner, R.D., Sventek, J.S.: In-line Service Measurements: An IPv6-based Framework for Traffic Evaluation and Network Operations. In: IEEE/IFIP Network Operations and Management Symposium NOMS 2004, Seoul, Korea (April 29-23, 2004)

[19] Pezaros, D.P., Sifalakis, M., Hutchison, D.: End-To-End Microflow Performance Measurement of IPv6 Traffic Over Diverse Wireless Topologies. In: Wireless Internet Conference WICON 2006, Boston, MA (August 2-5, 2006)

[20] Pezaros, D.P., Sifalakis, M., Mathy, L.: Fractal Analysis of Intraflow Unidirectional Delay over W-LAN and W-WAN Environments. In: Proceedings of the third International Workshop on Wireless Network Measurement (WiNMee 2007) and on Wireless Traffic Measurements and Modelling (WiTMeMo 2007), Limassol, Cyprus (April 20, 2007)

[21] Taqqu, M.S., Teverovsky, V., Willinger, W.: Estimators for long-range dependence: An empirical study. Fractals 3(4), 785–798 (1995)

[22] Willinger, W., Paxson, V., Taqqu, M.S.: Self-similarity and Heavy Tails: Structural Modelling of Network Traffic. In: Adler, R., Feldman, R., Taqqu, M.S. (eds.) A Practical Guide to Heavy Tails: Statistical Techniques and Applications, Birkhauser, Boston (1998)

[23] Willinger, W., Taqqu, M.S., Sherman, R., Wilson, D.V.: Self-Similarity Through High-Variability: Statistical Analysis of Ethernet LAN Traffic at the Source Level. IEEE/ACM Transactions on Networking 5(1), 71–86 (1997)

Effect of Transmission Opportunity Limit on Transmission Time Modeling in 802.11e

Nada Chendeb[1], Yacine Ghamri-Doudane[1], and Bachar El Hassan[2]

[1] Networks and Multimedia Systems Research Group (LRSM), Ecole Nationale Supérieure pour l'Informatique de l'Industrie et de l'Entreprise (ENSI.I.E.), 18 allée Jean Rostand, 91025 Evry, CEDEX – France

[2] Lebanese university, faculty of engineering, Rue al arz, el kobbeh, Tripoli, Lebanon

{chendeb,ghamri}@ensiie.fr, bachar_elhassan@ul.edu.lb

Abstract. Several analytical models have been developed for both 802.11 Distributed Coordinated Function (DCF) and 802.11e Enhanced Distributed Channel Access (EDCA). However, none of these models considers the 802.11e Contention Free Burst (CFB) mode which allows a given station to transmit more than one frame for each access to the channel. In order to develop a new and complete analytical model including all the differentiation parameters and able to be applied to all network conditions, we first need to know the time occupied by the transmission of a flow belonging to a given AC. The main objective of the current work is to analyze the TXOP bursting procedure and to propose a simple model allowing us to calculate the transmission time occupied by a particular AC when using the CFB mode. CFB performance analysis as well as the proposed model are discussed, demonstrated and validated by means of simulations.

Keywords: 802.11e, Analytical modeling, WLANs.

1 Introduction

IEEE 802.11 Medium Access Control (MAC) has become a defacto standard for wireless LANs. However, there are many inherent QoS limitations in the base standard. Consequently, a new standard IEEE 802.11e is specified [12]. It aims at supporting QoS by providing differentiated classes of service in the medium access control (MAC) layer and it also aims at enhancing the ability of all physical layers so that they can deliver time-critical multimedia traffic, in addition to traditional data packets. The concept of Transmission Opportunity (TXOP) or Contention Free Burst (CFB) is used in 802.11e; thus, neglecting it while drawing an analytical model for 802.11e is an important limitation for the model.

The remainder part of this paper is organized as follows. Related works and motivations are discussed in section 2. The main characteristics of IEEE 802.11e are briefly reviewed in Section 3. Effect of TXOP bursting on the global performance is analyzed in section 4. The proposed model for CFB feature and the corresponding

D. Medhi et al. (Eds.): IPOM 2007, LNCS 4786, pp. 156–167, 2007.

equations are presented in Section 5. Model validation and simulation results are analyzed and discussed in section 6. Finally, conclusions are drawn in Section 7.

2 Motivations

We are motivated to study and analyze this CFB feature because, at the level of our knowledge, and after comparing and examining all existing analytical models [1] – [11], we found that, there is not anyone taking into account the TXOP Limit differentiation parameter, that this fact will strongly affect the performance analysis.

Many of the researches undertaken up to now in the field of 802.11e performance modeling leave this parameter for future works while others assume that it is automatically present without really considering it. This can be considered as a clear gap in these models. Cited below are some statements that appear in some publications. For example, in [4], authors say: "...TXOP bursting extension is beyond the scope of this paper and is left for future work..." In [5], it was mentioned: "...In this paper, for simplicity, we only investigate the situation where a station transmits one data frame per TXOP transmission round..." In [8], the following affirmation is stated: "...Priority based on differentiated Transmission Opportunity (TXOP) limits is not treated explicitly in this paper..." and finally in [11], authors assume that "...One data frame per EDCF-TXOP is transmitted..." In [2], [3] and [7], TXOP bursting is not mentioned at all.

As a result, we are motivated to fill in this gap, and as a first step, we will model the effect of CFB mode on the transmission time, which itself affects directly the throughput and the access delay. A complete analytical model for 802.11e capable of calculating available throughput and mean access delay for all Access Categories (ACs) will be the subject of a future extension of this actual research.

3 IEEE 802.11e MAC Access Mechanisms

The QoS facility includes a Hybrid Coordination Function called HCF. The HCF uses both a contention-based channel access method, called the Enhanced Distributed Channel Access (EDCA) mechanism for contention based transfer and a controlled channel access, referred to as HCF Controlled Channel Access (HCCA) mechanism, for contention free transfer. IEEE 802.11e also defines a transmission opportunity (TXOP) limit as the time interval during which a particular station has the right to initiate transmissions. During a Transmission opportunity (TXOP), a station may be allowed to transmit multiple data frames from the same AC with a Short InterFrame Spacing (SIFS) between an ACK and the subsequent data frame [12]. This process is also referred to as Contention Free Burst (CFB) or TXOP bursting. Under HCF the basic unit of allocation of the right to transmit onto the Wireless Medium (WM) is the TXOP.

The EDCA mechanism defines four access categories (ACs). For each AC, an enhanced variant of the DCF, called an Enhanced Distributed Channel Access Function (EDCAF), contends for TXOPs using a set of EDCA parameters. These parameters are neither constant nor fixed by physical layer as with DCF, they are assigned either by a management entity or by a QoS AP (QAP). These differentiation parameters are basically: The idle duration time AIFSN (Arbitration Inter Frame Space Number), and the contention window minimum (aCWmin) and maximum (aCWmax) limits, from which the random backoff is computed.

A zero value for TXOP limit indicates that a single frame may be transmitted at any rate for each TXOP. Multiple frames may be transmitted in an acquired EDCA TXOP if there is more than one frame pending in the queue of the AC for which the channel has been acquired. If a QSTA has in its transmit queue an additional frame of the same AC as the one just transmitted, and if the duration of transmission of that frame plus any expected acknowledgement for that frame is less than the remaining medium occupancy timer value, then the QSTA may begin transmitting that frame at SIFS after completing preceding frame exchange sequence.

4 Effect of CFB on the Global Performance

Before presenting the proposed CFB model, let us first depict the effect of the CFB mode on the performances of the 802.11e EDCA access method. To show the effect of the TXOP bursting (CFB mode) on EDCA performance, a short simulation analysis is conducted. This one is realized using the ns-2 [13] enhanced with the TKN's 802.11e implementation [14].

4.1 Simulation Topology and Parameters

Node topology of the simulation consists of five different wireless QSTAs, contending for channel access, one AP and one wired station. All wireless nodes send their data to the wired node via the AP. These 5 wireless nodes and the AP are all situated in the same radio range and distributed as shown in Fig. 1. Each QSTA uses all four ACs. Poisson distributed traffic, consisting of 800-bytes packets, was generated at equal rate to each AC. This simulation scenario will be run with different Poisson arrival rates going from 50 Kb/s to 2100 Kb/s for each AC at each QSTA, the total arrival rate in the simulated network varies from 1000 Kb/s (non saturated) to 42000 Kb/s (fully saturated).

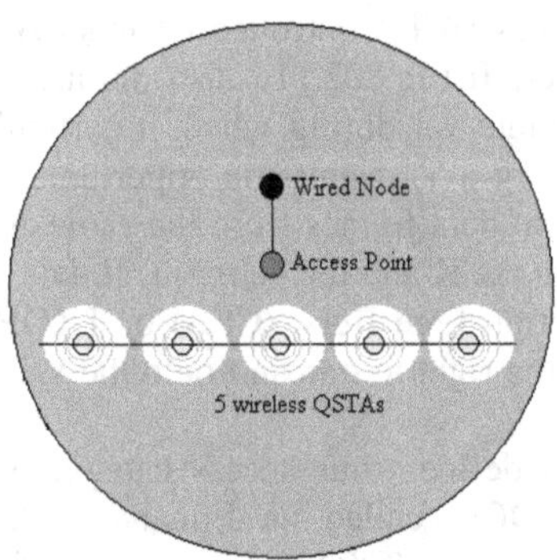

Fig. 1. Simulation topology

4.2 Impact of CFB Mode on EDCA Performance

First of all, and in order to show the prioritized access in EDCA, let us analyze for each AC the number of accesses to the wireless medium during the simulation time. Fig. 2 shows the percentage of the number of accesses to the channel with respect to the arrival rate. This figure shows clearly that for low arrival rates (until 200 kb/s), all ACs have identical percentage of access to the channel, the network can serve equally all the ACs without problems. When the load rate increases, the differentiation between ACs becomes very clear. AC_0 gets 60% of the total access, while AC_3 cannot exceed 5% of the total access.

Now, to analyze the effect of the TXOP limit differentiation parameter, we ran the simulation in two different modes. In the first one, CFB was activated, and in the second one, it was deactivated. Fig. 3, 4 and 5 show the effect of this differentiation parameter on the performance metrics such as throughput and delay.

It can be clearly seen in Fig. 3 that the higher the TXOP is the higher will be the throughput improvement for the corresponding ACs. Furthermore, in CFB, all ACs gain more throughputs as compared to the case without CFB. So the total throughput in the network becomes higher, and the medium capacity is well used. The positive influence of CFB appears especially for AC_1 which represents the video traffic. This can be explained by the fact that CFB allows video application to transmit the highest number of frames in a burst in place of one frame, and this is the main objective of 802.11e, that is to allow multimedia traffic and QoS based application to be transmitted correctly in WLANs.

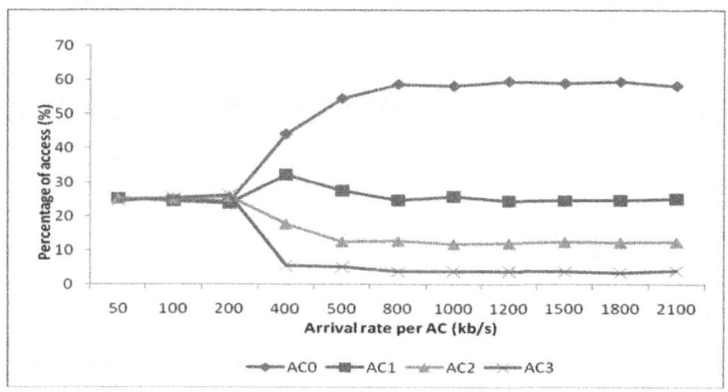

Fig. 2. Percentage of number of accesses with respect to arrival rate

The effect of CFB appears also on the mean access delay of all ACs. Since the mean access delay for AC_0 and AC_1 is very small as compared to those of AC_2 and AC_3, we draw here two sets of curves, the first set of curves (Fig. 4) shows the small values related to AC_0 and AC_1 while the second set of curves (Fig. 5) shows the values obtained for AC_2 and AC_3.

Fig. 4 and Fig. 5 show that the values of access delay we got in CFB mode in the majority of load rates are greater than those we got without CFB. To explain this result, we can say that the burst transmission procedure, forces contending traffics to

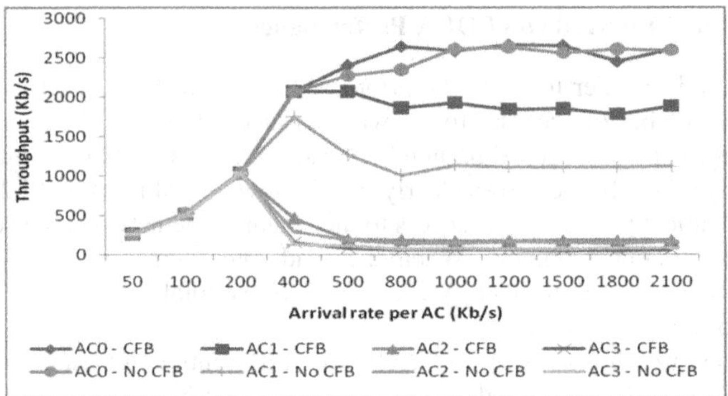

Fig. 3. Effect of CFB on the available throughput

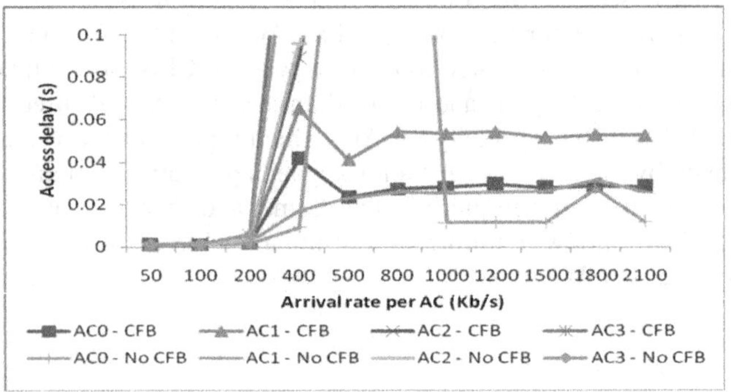

Fig. 4. Effect of CFB on the access delay (focus on AC_0 and AC_1)

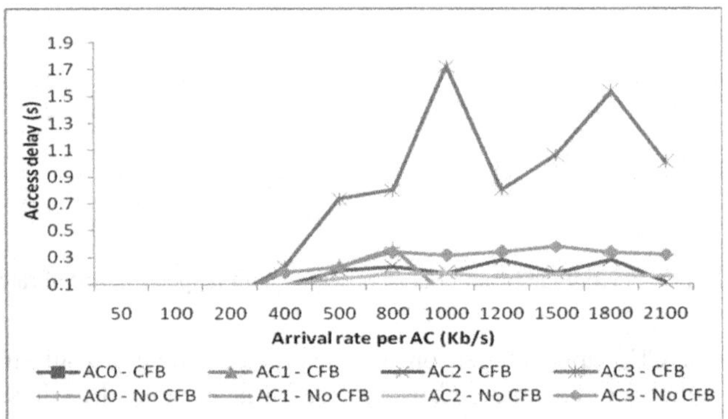

Fig. 5. Effect of CFB on the access delay (focus on AC_2 and AC_3)

wait an additional time, thus increasing the total mean access delay for all ACs. On the other hand, AC_0 which represents the voice traffic suffers from a hard access delay where CFB is not active. This is especially seen in the saturation limit (between

400 and 1000 Kb/s). In this interval, AC_0 (voice) has an access delay greater than that of AC_2 (best effort). As a conclusion, the differentiation in terms of access delay is better with CFB.

So, we can affirm that the global performance in 802.11e is clearly different (enhanced) when using CFB compared to when not using it. Thus, an accurate modeling for this feature seems important.

5 Transmission Time Modeling in CFB Mode

Almost all of the analytical models done for the 802.11e EDCA Access mechanism have been constructed in saturation conditions; this means that they suppose each AC queue always has data to be transmitted. Moreover, in these models, it is assumed that once an AC accesses the wireless medium, it transmits one and only one data frame. This assumption taken by all of these models [1] – [11] facilitates the calculation and makes the models simpler concerning data transmission time. However, such simplification leads to a non accurate estimation of available bandwidth and access delays while CFB mode is used. This consideration is highlighted by the results we obtained in section 4. Additionally, in most usual cases, TXOP is not set to null for all ACs, and it is considered as an important differentiation parameter between different ACs, giving some ACs more priority to transmit additional data as compared to other ACs. Here, we aim at calculating the transmission time occupied by a particular Access Category (AC_i, i = 0...3) when accessing the channel. We consider that the CFB mode is running in the network.

Let P be the payload transmission time for a data frame and H the time needed to transmit the header[1]. According to the access mode, the transmission time for a single data frame may vary. This time is named T_{s1}. It is given in (1) and (2).
In RTS/CTS access mode:

$$T_{s1} = RTS + \delta + SIFS + CTS + \delta + SIFS + H + P + \delta + SIFS + ACK + \delta \,. \tag{1}$$

In basic access mode:

$$T_{s1} = H + P + \delta + SIFS + ACK + \delta \,. \tag{2}$$

Where δ is the propagation delay between nodes.

The number of frames that can be transmitted by AC_i in a given access to the wireless medium is limited by its $TXOP_i$. Since frames transmitted during a $TXOP_i$ bursting are separated by a SIFS time, we may conclude that during this $TXOP_i$, AC_i can transmit a number of frames equal to:

$$N_{TXOPi} = \left\lceil \frac{TXOP_i}{T_{s1} + SIFS} \right\rceil \text{ (Upper integer value).} \tag{3}$$

Note that for $TXOP_i = 0$, we have $NTXOP_i = 1$.

---

[1] H is separated from P in the formulation, because the data rate used for these two quantities is different.

N_{TXOPi} constitutes the maximum number of frames that an AC_i can transmit in its access to the channel. However, the queue of a given ACi may contain, in some conditions of the arrival rate, a number of frames less than N_{TXOPi}, and in other conditions a greater number. In the first case, it transmits only the existing frames in its queue while in the second case, it transmits N_{TXOPi} frames. In the second case, known as the saturation case, the number of frames transmitted is known and can be calculated by equation (3). But to resolve the problem in the first case (non saturation case) we are led to calculate the queue length of the AC_i.

In our model, it is supposed that we have an infinite M/G/1 queue with a Poisson arrival process of rate λ to model the arrival of packets at the MAC buffer of an AC. Even though the Poisson assumption may not be realistic, it provides insightful results and allows the model to be tractable.
According to this assumption, the average number of packets in the system is equal to:

$$N_{Avi} = \frac{\rho_i}{1-\rho_i} \text{ with } \rho_i = \lambda_i \overline{D}_i \quad 0 \le N_{Avi} \le \infty \tag{4}$$

$\overline{D}_i$ is the mean access delay of AC_i to the channel. Obtaining $\overline{D}_i$ is not a simple matter. Indeed, we need to construct a complete analytical model capable of calculating it. On the other hand, to be accurate, such complete analytical model needs a model for transmission time under the CFB mode. As stated previously, the complete accurate analytical modeling of EDCA will be the target of our future research. Thus, in order to validate transmission time model under the CFB mode, we use delay values obtained from ns-2 simulation.

Finally, we can say that, in general, the number of frames transmitted in each access to the channel is:

$$N_{Transi} = \min (N_{Avi}, N_{TXOPi}) \tag{5}$$

And the transmission time in a transmission cycle for all ACs can be given by the following equation:

$$T_{si} = N_{Transi_i}(T_{s1} + SIFS) \quad 0 \le i \le 3 \tag{6}$$

6 Model Validation and Analysis

In order to validate the proposed model and to show its usefulness, a set of simulations, using the CFB mode, are realized. The network topology used to validate the model is the same as that described in section 4. The selected physical protocol for validations is 802.11b. 802.11e parameters, such as AIFSN, CWmin, CWmax and TXOP limit, are overridden by the use of 802.11e EDCA default parameters [12].

Using the 802.11b parameters, by calculation of the equation (2), we get $T_{s1} = 1066$ µs. Application of equation (3) gives the following results: $N_{TXOP0} = [3.03] = 3$, $N_{TXOP1} = [5.59] = 5$. The TXOP limit for AC_2 and AC_3 being equal to 0, only one frame will be sent per-access. $N_{TXOP2} = N_{TXOP3} = 1$.

6.1 Model Validation

Figures 6 to 8 prove the accuracy of the proposed equations in calculating the transmission time.

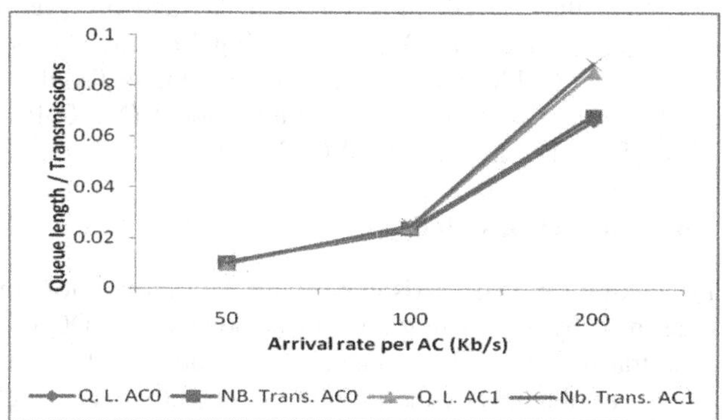

Fig. 6. Average queue length versus average number of transmissions, before the saturation limit (AC_0 and AC_1)

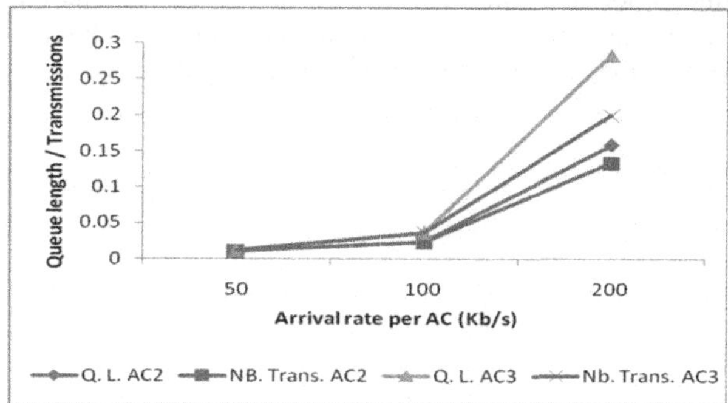

Fig. 7. Average queue length versus average number of transmissions, before the saturation limit (AC_2 and AC_3)

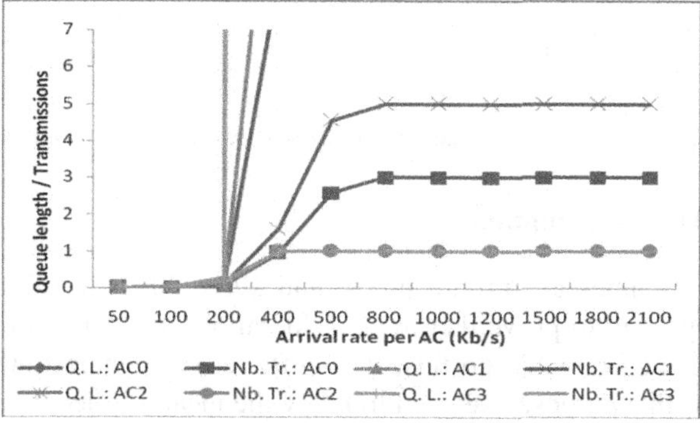

Fig. 8. Average queue length versus average number of transmissions, for different load rates and for all ACs

In Fig. 6 and 7, it may be noticed that before the saturation, the average queue length (Q. L.) coincides with the average number of transmissions (Nb. Tr.). This number is less than 1 on average because most of time, the queue is empty and there is no frame to transmit. While approaching the saturation limit, the queue length starts reaching values greater than TXOP limit for each AC and the average number of transmission starts reaching the NTXOPi (Fig. 8). Fig.8 clearly shows that NTXOPi obtained in the simulation (NTXOP0 = 3, NTXOP1 = 5 and N TXOP2 = N TXOP3 = 1) is equal to that obtained by through our simple model (NTXOP0 = [3.03] = 3, NTXOP1 = [5.59] = 5 and N TXOP2 = N TXOP3 = 1).

6.2 Comparison with Previous Models

In Fig. 9, we are looking for a comparison between the transmission time while CFB is disabled (as assumed by all existing analytical models for EDCA), which do not consider CFB, the transmission time as calculated by our simple model, and finally the transmission time as obtained by the simulation. The previous models assume that one data frame is transmitted per transmission round, so the calculation of the transmission time they consider and which is get in the simulation in case of our scenario gives a value likely equal to 1ms whatever is the load rate. In our model, transmission time is not stable all the time; it depends on the load rate. Fig. 9 shows that the average transmission time we got through equations is closely identical to that we got by the simulation, especially in the saturation conditions.

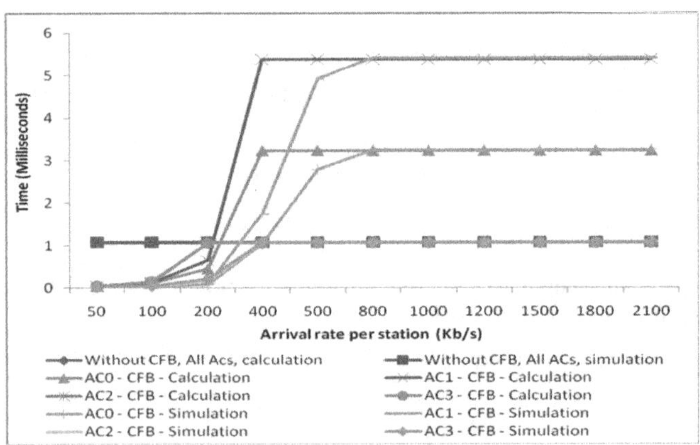

Fig. 9. Transmission time, previous models, our model, simulation

6.3 Effect of Used Assumption

As our model assumes an infinite queue and since our simulations use a finite queue, it is important to prove that the consideration of a Poisson process with an M/G/1 queue is accurate. To do this, it is sufficient to demonstrate that equation (4) is valid. For this purpose, we need to know the mean service time which is here equal to the mean access delay to the wireless medium. The mean access delay for each AC is obtained from the NS-2 simulations, and it is drawn in Fig. 4 and

Fig. 5. In Fig. 10 we can distinguish two different cases to analyze: Before the saturation limit (<500 Kb/s) and within the saturation limit (between 500 and 1000 Kb/s). The zone situated after the saturation limit (saturation conditions) was not drawn here, because in this regime, equation (4) is not used and the transmission time depends only on the TXOP limit. Before the saturation limit, there is a very slight difference between the average queue length we had by the simulation and that we had through equation (4). As this difference is not really noticeable, we can consider that equation (4) approximates this case very well and thus accurately models it. Within the saturation limit, results are not the same; it is observable from Fig. 10 that there is an important difference between calculations and simulations especially for AC_2 and AC_3. Note however that this difference does not really alter our model. Indeed, even if equation (4) does not model very well this case, the fact that it highly over-estimates the number of packets in queue is mitigated by equation (5). Actually, we can easily notice that values resulting from calculations are greater than the corresponding NTXOP for these two ACs, and as it is noticed in equation (5), the minimum value will be considered as the number of transmitted frames. Thanks to this mitigation, the number of transmitted packets per TXOP (cf. equation (5)) remains accurate. As a conclusion, we can say that this model is globally accurate, and the error produced in the saturation limit is mitigated in the global performance of the model. After the saturation limit, the calculations and simulations produce very high values, and in this regime N_{TXOP} will be considered all the time.

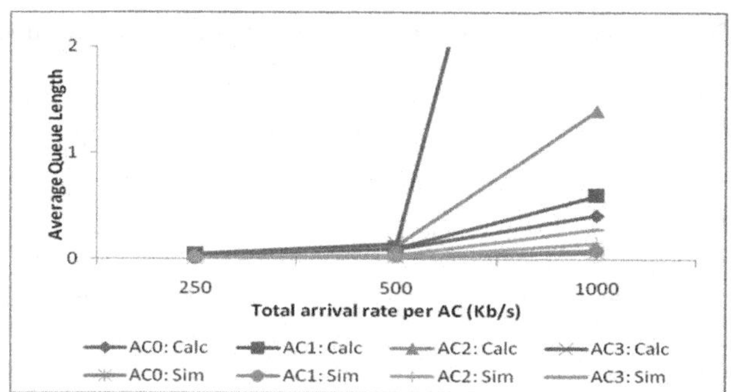

Fig. 10. Average queue length: analytical model versus simulation

6.4 Queue Length at Each Transmission Attempt

Finally, to track the state of the queues at each access to the channel, Fig. 11, 12 and 13 show the number of frames in the queue of AC_1 and the number of transmitted frames for each access and for three network conditions. Before the saturation limit, the queue size is always equal to the frame transmitted because this number is less than N_{TXOP1} (having 0 and 1 frame almost all the time, this leads to an average number of transmissions less than 1). Within the saturation limit, we have almost all

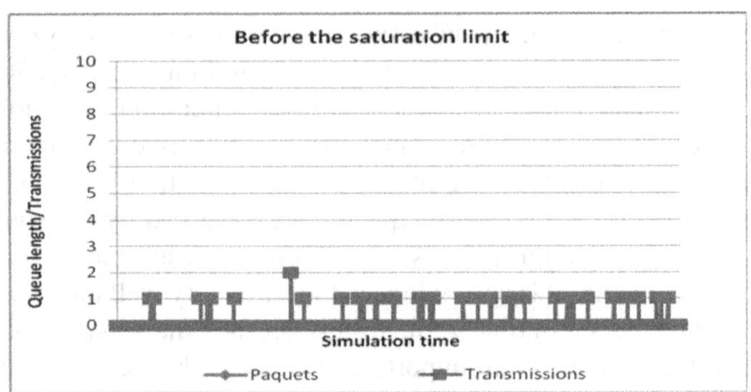

Fig. 11. Queue length and number of transmission before the saturation limit

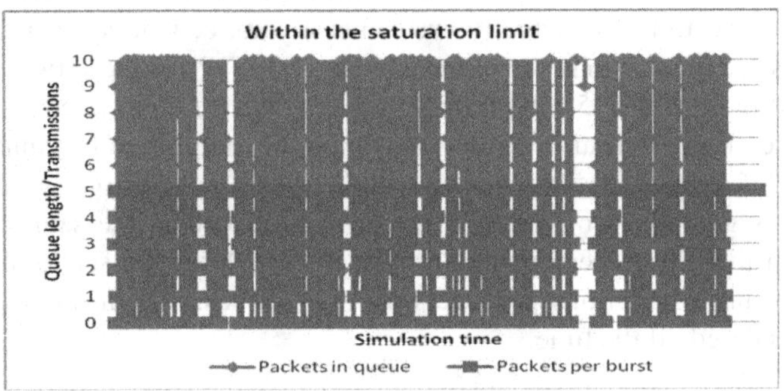

Fig. 12. Queue length and number of transmission within the saturation limit

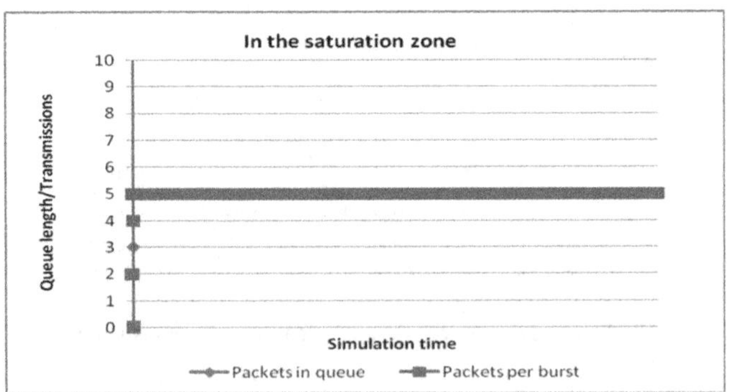

Fig. 13. Queue length and number of transmission in saturation conditions

possibilities (0, 1, 2, 3, 4, 5) and in saturation conditions, the queue length is very large and the number of transmissions is always equal to $N_{TXOP1} = 5$.

According to the results obtained here, we can say that the simple model we proposed for the calculation of the transmission time is accurate.

7 Conclusions

In this paper, we proposed a simple model to calculate the transmission time in CFB mode. As far as we know, this is the first time this feature is modeled.

After having highlighted the influence of the CFB mode on 802.11e performances and the importance of taking it into account when modeling the 802.11e EDCA function, we demonstrated by means of simulations that the simple CFB model we proposed is accurate. As a next step to this work, we are currently working on the integration of this model into a complete model capable of calculating the available bandwidth and the access delay for all ACs in 802.11e.Such complete model, once implemented in wireless nodes, will permit the calculation of the available bandwidth and the access delay for each AC and so allowing to perform admission control for any new traffic belonging to a given AC.

References

1. Bianchi, G.: Performance analysis of the IEEE 802.11 Distributed Coordination Function. IEEE Journal on Selected Areas in Communications 18, 535–547 (2000)
2. Tao, Z., Panwar, S.: An Analytical Model for the IEEE 802.11e Enhanced Distributed Coordination Function. In: LANMAN 2004, San Francisco, California, USA (April 2004)
3. Zhu, H., Chlamtac, I.: An analytical model for IEEE 802.11e EDCF differential services. In: ICCCN 2003, Dallas, Texas, USA (October 2003)
4. Robinson, J.W., Randhawa, T.S.: Saturation throughput Analysis of IEEE 802.11e Enhanced Distributed Coordination Function. IEEE Journal on Selected Areas in Commuications 22(5) (June 2004)
5. Kong, Z., Tsang, D.H.K., Bensaou, B., Gao, D.: Performance Analysis of IEEE 802.11e Contention-Based Channel Access. IEEE Journal on Selected Areas in Communications 22(10) (December 2004)
6. Wang, K.C., Ramanthan, P.: End-to-End Throughput and Delay Assurances in Multihop Wireless Hotspots. In: WMASH 2003. 1st ACM Workshop on Wireless Mobile Applications and Services on WLAN Hotspots, San Diego, California, USA (September 2003)
7. Chen, Y., Zeng, Q.-A., Agrawal, D.P.: Performance of MAC Protocol in Ad Hoc Networks. In: CNDS 2003. Communication Networks and Distributed Systems Modeling and Simulation Conference, Orlando, Florida, USA (January 2003)
8. Engelstad, P.E., Østerbø, O.N.: Queueing Delay Analysis of IEEE 802.11e EDCA. In: Wireless On demand Network Systems and Services conference, Les Ménuires, France (January 2006)
9. Chen, Y., Zeng, Q.-A., Agrawal, D.P.: Performance of MAC Protocol in Ad Hoc Networks. In: CNDS 2003. Communication Networks and Distributed Systems Modeling and Simulation Conference, Orlando, Florida, USA (January 2003)
10. Engelstad, P.E., Østerbø, O.N.: Analysis of the Total Delay of IEEE 802.11e EDCA and 802.11 DCF. In: ICC 2006, Istanbul, Turkey (June 2006)
11. Mangold, S., Sunghyun Choi Hiertz, G.R., Klein, O., Walke, B.: Analysis of IEEE 802.11e for QOS Support in Wireless LANS. Wireless Communications 10(6), 40–50 (2003)
12. Wireless medium access control and physical layer specifications: Medium Access Control Quality of Service Enhancements, IEEE P802.11e/D13.0 (January 2005)
13. The Network Simulator - ns-2, http://www.isi.edu/nsnam/ns/
14. IEEE 802.11e implementation for ns-2, http://www.tkn.tu-berlin.de/research/802.11e_ns2/

Cognitive Network Management with Reinforcement Learning for Wireless Mesh Networks

Minsoo Lee, Dan Marconett, Xiaohui Ye, and S.J. Ben Yoo

Dept. Electrical and Computer Engineering, University of California, Davis
Davis, CA 95616 USA
{msolee,dmarconett,xye,sbyoo}@ucdavis.edu

Abstract. We present a framework of cognitive network management by means of an autonomic reconfiguration scheme. We propose a network architecture that enables intelligent services to meet QoS requirements, by adding autonomous intelligence, based on reinforcement learning, to the network management agents. The management system is shown to be better able to reconfigure its policy strategy around areas of interest and adapt to changes. We present preliminary simulation results showing our autonomous reconfiguration approach successfully improves the performance of the original AODV protocol in a heterogeneous network environment.

Keywords: Cognitive networks, autonomic management, wireless mesh networks.

1 Introduction

Due to the fact that current network management systems have limited ability to adapt to dynamic network conditions, providing intelligent control mechanisms for globally optimal performance is a necessary task. With respect to the current generation of management technology, there are limitations to the performance enhancements they can provide, due to the lack of contextual awareness of network conditions. This limitation results in obstacles to efficient optimization of network performance. In the case of routing in wireless mesh networks, battery-powered devices create challenging problems in terms of prolonging the autonomous lifetime of the network. In designing intelligent routing protocols, the various features of such multi-hop wireless networks, such as power-limited sensor networks, lead to a set of optimization problems in routing path length, load balancing, consistent link management, and aggregation [1]. Most existing routing techniques are designed to optimize one of these goals. However, these factors are usually in competition, and influence the routing performance in a complex way. Clearly, solving the optimization goals independently will not lead to an optimal solution. Rather, one needs to consider all the optimization concerns in the aggregate, when addressing the issue of intelligent network-layer management.

A possible solution to these multi-faceted management issues is presented in the form of cognitive networks. Cognitive wireless networks are capable of reconfiguring

D. Medhi et al. (Eds.): IPOM 2007, LNCS 4786, pp. 168–179, 2007.

their infrastructure, based upon experience, in order to adapt to continuously changing network environments [2]. Cognitive networks [3] are seen as a major facilitator of future heterogeneous internetworking and management, capable of continuously adapting to fluid network characteristics as well as application-layer QoS requirements. Typically, Machine Learning techniques, such as Q-learning [4], can expedite the ability for autonomous network management agents to adapt, self-configure, and self-manage. Specifically, autonomic self-configuration of heterogeneous network systems will have cross-layer ramifications, from the physical (PHY), Medium Access Control (MAC), network, transport layers to the middleware, presentation and application layers. Therefore, cross-layer design [5-8] approaches are critical for the efficient utilization of limited resources, to better provide QoS guarantees for end-users, in future wireless mesh networks. This management approach can overcome the potential scope limitations of network management in heterogeneous wireless networks, by allowing autonomous agents to observe and adapt, in order to dynamically optimize management policy over time. Our strategy is to augment the routing protocol in wireless multi-hop networks with a distributed Q-learning mechanism, to ensure that Service Level Agreements (SLA) and the packet delivery ratio can be maintained at desired levels, while minimizing additional management overhead. As such, the concerns which need to be addressed are threefold:

1) What contextual information needs to be exchanged between layers?
2) Which network parameters should be subject to dynamic reconfiguration?
3) When should the results of Q-learning reinforcement be utilized reconfigure relevant network parameters?

We propose preliminary work which improves network performance, by providing an underlying network management system, maintained by the proposed Q-learning based autonomic reconfiguration. Specifically, we present a novel, reconfigurable wireless mesh network routing management architecture, which enables network nodes to efficiently learn an optimal routing policy, enhancing the packet delivery ratio and QoS support. Preliminary simulation results show that our autonomous cross-layer reconfiguration successfully improves the scalability of the original AODV protocol, in a heterogeneous network environment. The remainder of the paper is organized as follows: Section 2 presents a belief survey of related work. Section 3 gives an overview of our network architecture with reinforcement learning techniques for autonomic reconfiguration. Section 4 describes in detail our Q-learning based management mechanism. Section 5 describes the simulation scenarios used for performance comparison. Section 6 discusses the derived simulation results. Finally, we discuss future work and conclude in Section 7.

2 Related Work

2.1 Cross-Layer Approaches for Intelligent Network Management

The realm of network management encompasses many concerns, including issues such as IP configuration, security, and network resource provisioning. While these concerns are not unique to wireless mesh, they are exacerbated by nodal mobility, dynamic network membership, and unstable links [9]. Depending on the speed of the Mobile Nodes (MNs), mobility can generally be classified into three categories of increasing speed: static, low mobility, and high mobility. The management of a network should be able to take into account any of these three cases and their associated performance implications. In the case of low mobility, the steady-state performance should be optimized and incidental updates (e.g., for route discovery) can be allowed to consume more resources; whereas in the high mobility case, resource consumption and delay due to route maintenance and updating are more important factors [10].

However, centralized network management architectures fail in heterogeneous mesh internetworking scenarios. Recently, the distributed decision making scheme [11] has been introduced to address the aforementioned concerns. In this proposed scheme, nodes may only be aware of the status of their one-hop neighbors, and have no sense of the size and extent of the network. Finding a mechanism that can manage the particular challenges associated with distributed decision making in mesh networks is certainly non-trivial. To cope with the demands of cross-layer management [5-8], this paper presents a solution which deploys the cross-layer mechanism between the application layer and the network layer in a MN, for exchanging end-to-end delay metrics.

2.2 Cross-Layer Autonomic Routing

There are several proposed "network-aware" routing schemes in the current literature [5-9] for multi-hop wireless networks. However, before reviewing some relevant routing approaches in wireless mesh, it is necessary to give a brief overview of Machine Learning in general, and Q-Learning in particular. Reinforcement Learning is a form of Machine Learning, characterized by the formulation of policy to achieve specific goals. Reinforcement Learning problems are typically modeled by means of *Markov Decision Processes* (MDPs) [4]. The model is comprised of the set of potential environment states, S; the set of possible actions, A; the designated reward function, $R: S \times A \rightarrow R'$; and the policy which determines state transition, $P: S \times A \rightarrow \pi(S)$. The set $\pi(S)$ represents the set of functions over set S, which specify the actions required to transition by means of action a □ A from state s to state s'. The instantaneous reward (S, A) is a result of taking action a while in a particular state.

The particular form of Reinforcement Learning that our network management mechanism employs is referred to as Q-Learning [4]. As a model-free reinforcement learning technique, Q-Learning is ideally suited for optimization in dynamic network environments. Model-free techniques do not require any explicit representation of the environment in which policy updates are formulated, and as such, can solely address parametric optimization to maximize long-term reward. In particular, Q-Learning employs an exponential weighted moving average calculation, to not only take note of recent policy success/failure as feedback, but also take into account the weighted average of past values observed, referred to as 'Q-values.' Q-value computation is performed via the following equation

$$Q(s',a') = (1-\alpha)Q(s,a) + \alpha \cdot r \quad (1)$$

Q(s,a) refers to the Q-value computed based upon state '*s*' and action '*a*.' In a networking environment, the state could be represented by the load on all links in the current network, and the action could correspond to the routing decision which led to the current levels of link utilization. Parameter '*r*' represents the instantaneous reward value which is derived from a measurement of the current environment based upon the current policy. The variable 'α' is referred to as the learning rate, or the weight we assign to the current observation, between 0 and 1. Note that in this implementation, the assigned learning rate value also determines the weight of previous Q-values in terms of the '(1-α)' coefficient. *Q(s',a')* is the new Q-value computed corresponding to the current action *a'* and the new resulting state *s'*. By measuring Q-values over time, the relative fitness of a particular policy may be ascertained, and obligatory iterative modifications to this policy can be performed.

2.3 Current Cross-Layer Adaptive Routing Protocols

Fundamental to adaptive routing is the identification of relevant performance metrics. The modification of the routing protocol itself is determined by the observation and analysis of these metrics of interest. Quality-Aware Routing (QAR) [20] is a non-RL routing scheme which attempts to address this issue by using two new parameters, modified expected number of transmissions (mETX) and effective number of transmissions (ENT), to reduce the packet loss rate in wireless mesh routing [20]. The modified expected number of transmissions is obtained by changing the previously used ETX metric to take into account the variability of radio-channel conditions, rather than the average channel behavior. A more responsive ETX is generated by calculating the combined expected value of both the average and variance of the packet error probability. The effective number of transmissions considers both the overall network capacity, as well as the higher layer protocol end-to-end loss-rate concerns. The value of ENT is defined as the sum of the average bit error probability of the channel and a so-called strictness coefficient times the packet error variance for that channel, referred to as the log ENT. Specifically, a given link is said to satisfy the ENT constraints if the calculated value is below the log value of the threshold of retransmits for the underlying link-layer protocol.

Routing in QAR is performed by modifying existing reactive routing protocols (e.g. DSR, AODV) to utilize mETX and ENT values during path selection. Probe packets are periodically sent to measure the ETX of a given link. Each node sends ten probe packets to sample the link loss rate, which is estimated by using an exponentially weighted moving average (EWMA) filter. The weight of the EWMA is halved every five sample iterations. The authors specify two routing approaches, the first being the periodic calculation of mETX for each link and assignment of the mETX value as the corresponding link cost. Standard cost minimization routing is then employed. The second approach utilizes the *log(ENT)* values, compared to *log(M)*, where *M* is the retransmit limit of the link-layer protocol. An infinite cost is assigned to links which have a *log(ENT)* value greater than *log(M)*. The ETX value is assigned as the link cost to all other links.

A further enhancement to RL-based routing can be found in the *AdaR* routing protocol [1]. AdaR uses the model-free reinforcement lear`ning technique of Least Squares Policy Iteration (LSPI) to iteratively approximate reward values of chosen routing policies, much the way Q-Learning does. However, LSPI has been found to converge must faster upon a optimal route policy than traditional Q-Learning based routing. LSPI approximates the Q-values for a particular routing policy decision. In the AdaR routing scheme, the current state *s* represents a particular ad hoc node, and state *s'* corresponds to its chosen neighbor to which the current packet is forwarded, via action *a*. Though the protocol is inspired by routing in ad hoc sensor networks, it is entirely appropriate to consider its implementation for volatile and time-varying hybrid wireless mesh networks. The protocol considers four metrics of interest for QoS routing in each state-action pair (s,a):

1) $d(s, a)$: the hop count differential between *s* and *s'* to the destination.
2) $e(s, a)$: the residual energy of node *s'*.
3) $c(s, a)$: the number of routing paths which incorporate node *s'*, which is determined by network state sampling.
4) $l(s, a)$: the estimated link reliability between nodes *s* and *s'*.

Basis functions for the (s, a) pairs are of the form

$$\phi(s, a) = \{d(s, a), e(s, a), c(s, a), l(s, a)\} \tag{2}$$

and the values for each are normalized to a range of [-1,1]. LPSI is used to update the weight values of the linear functions, based upon the network samplings. Weight updates are used to modify the current policy iteratively.

2.4 Q-Routing Versus Q-Management

In the above described routing approaches, reinforcement learning, Q-Learning in particular, is used to determine the actual routing paths in the multi-hop wireless network. This solution is productive if one is merely concerned with the cross-layer design of a specific routing protocol. However, it is quite clear from the literature that no single routing protocol will facilitate robust performance in all scenarios [11]. Mobile and static wireless mesh networks have fundamentally different needs in terms of route management and routing traffic generation, as well as the support of application-layer QoS needs. When concerned with the creation of a general mesh management framework, which caters to the needs of a dynamic network, a far better approach is to have access to a dynamic proactive and reactive routing protocol for relatively static and mobile mesh networks respectively. Depending on network characteristics, the appropriate protocol can be implemented and augmented in real-time by the intelligent network management agent(s) responsible for self-management and reconfiguration. The creation of intelligent management mechanisms to dynamically reconfigure existing flat routing protocols is the first step to realize this goal, and we discuss our implementation and initial results in the subsequent sections.

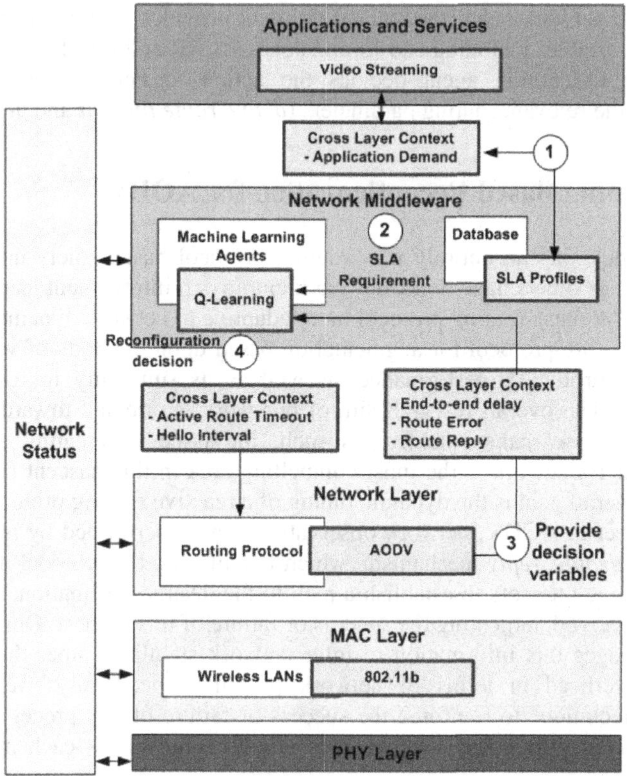

Fig. 1. Cross-layer desgin for distributed decision making in a mobile node

3 Cross-Layer Autonomic Network Architecture

Initial proposals on the implementation of cross-layer management interaction are discussed in the current literature. These approaches can be divided into three categories [8]: direct communication between layers, a shared database across the layers, and novel contextual abstractions. We propose cross-layer interactions among layers by a shared network status module, supporting vertical communications among the layer s, by acting as a repository for information collected by network protocols. Each layer-specific module can access the shared network status module to exchange its data and interact. Figure 1 is an illustration of our cross-layer design approach, and visually conveys how our Q-learning mechanism interacts with other reconfigurable modules from a cross-layer perspective. The following steps describe the detailed procedure of our cognitive network management.

Step 1: The SLA management module in the middleware layer gathers application demands and set SLA requirements.
Step 2: The Q-learning agent in the middleware layer receives the SLA requirements and sets the level of SLA compliance.

Step 3: At the network layer, the AODV protocol provides the Q-learning agent with the decision variables including end-to-end delay, RERR and RREP.
Step 4: The Q-learning agent decides the action to enhance performance and reconfigures the relevant routing parameters (*active route timeout* and *hello interval*).

4 Q-Learning Based Reconfiguration for AODV

Rather than devising an entirely new routing protocol based solely upon link-layer measurement, as others have done, our work employs reinforcement learning to make and existing flat mesh routing protocol more adaptive to network dynamics. We chose the AODV routing protocol for augmentation based upon the body of work available regarding the protocol's performance, as well as its suitability for dynamic mesh environments. The overall research aim of our work is oriented towards cognitively intelligent network management. As such, the issue of creating an intelligent network-layer framework is the most compelling area in this nascent field. The first step in this overall goal is the dynamic tuning of a reactive routing protocol, AODV.

With respect to AODV, network observation can be performed by monitoring the *route_request/route_reply* mechanism, which is built into the protocol itself. When a particular r node attempts to establish a path to the desired destination, a *route_reply* message is received, indicating the success or failure of this request. Our management scheme leverages this information to infer network stability, rather than generating additional overhead in terms of network probing. Specifically, we employ the Q-Learning technique to reinforce the success or failure of this process, and modify our management policy accordingly. This modification entails each node retaining two Q-values:

- Q[1]: *penalty Q-value for non-successful route requests*
- Q[2]: *reward Q-value for successful route requests*

The iterative updating of Q-values is determined by the following equation:

$$Q[i] = (1-\alpha)Q[i] + \alpha \cdot r \tag{3}$$

For our experimentation, we chose the end-to-end delay of the route reply packet as the penalty of failure, and the inverse of the delay as the reward for success, corresponding to the reward/penalty function value '*r*.' Q[1] denotes the Q value for unstable network status that makes the node take the action of decreasing *active route timeout* and *hello interval*. Q[2] presents the stability of the network that will make the node take the action of increasing *active route timeout* and *hello interval*. Each node makes its reconfiguration decision based on the local routing information, represented as the two Q values which estimate the quality of the alternative actions. These values are updated each time the node receives a RREP packet in the following manner:

- Reward: $r = n\ (1\ /\ (ETE_t\ /\ ETE_{max}\ )\ )$
  1. When a ROUTE REPLY packet reaches the source and there is a path from the source to destination.
- Penalty : $r = n\ (ETE_t\ /\ ETE_{max}\ )$

2a. When a ROUTE ERROR packet is generated as the route is broken.
2b. When a ROUTE REPLY packet reaches the source and there is no path from the source to destination.

where ETE_t is the current end-to-end delay, ETE_{max} is the maximum end-to-end delay that complies with the SLA agreement and the parameter 'n' denotes the normalization constant.

5 Simulations

The performance of our network system has been evaluated with the OPNET simulation tool [18]. The simulation network is defined as a flat terrain of 3000 × 3000 m with 25 MNs, 2 MANET gateways, 1 mobile wireless video streaming server and 2 Ethernet video streaming servers. The mobile wireless video streaming server and MNs were equipped with the wireless network interface. At the physical and data link layers, the 802.11b standard was used for the analysis. The 802.11b standard was selected, as it is readily available in OPNET and provides the capability of defining different transmission data rates for the mobile nodes. It should be noted that the main purpose of the simulation scenarios was to provide a framework to compare the performance of our Q-learning based reconfiguration in AODV versus the standard AODV protocol.

The traffic model used to gather the simulation results consists of three constant bit rate (CBR) sources. The traffics are defined as follows:

- Traffic #1 Mesh-to-Mesh: 300 kbps downloading video streaming from Wireless Video Streaming Server to mobile_node_1.
- Traffic #2 Mesh-to-Ethernet: 300 kbps uploading video streaming from mobile_node_2 to Ethernet Video Streaming Server 1.
- Traffic #3 Ethernet-to-Mesh: 300 kbps downloading video streaming from Ethernet Video Streaming Server 2 to mobile_node_3.

In terms of SLA compliance, the maximum end-to-end delay, ETE_{max}, was $3ms$ which complies with the SLA agreement.

In this analysis, node mobility is assumed to be random (i.e., independently selected by each node using a uniform distribution) movement rather. The mobile nodes are assigned a maximum speed of 15 m/s and each mobile node changes its location within the network based on the "random waypoint" model. In order to calculate the impact of high mobility on the protocol overhead, pause-time is assigned 0 s. It should be noted that a pause-time of 0 s represents the worst case scenario as the mobile nodes are constantly moving during the simulation. In the training episode, from the beginning of simulation to 150 s, each node randomly chooses actions decreasing or increasing *active route timeout* and *hello interval.*

6 Performance Evaluation

The performance of our Q-learning based reconfiguration was evaluated in terms of packet delivery ratio, end-to-end delay, SLA compliance, and the overhead load in the network due to control messages generated by the AODV routing mechanism. The performance results derived under our Q-learning based reconfiguration were compared with the performance results derived under the flat AODV routing mechanism. Table 1 displays the summary of our OPNET simulation results.

6.1 Control Overhead and Route Errors

The control overhead is measured in terms of the number of control messages generated by the routing algorithm. Figure 4 illustrates the number of routing control messages generated or relayed in the network. On average, our Q-learning based reconfiguration achieves a reduction of control messages equivalent to 55.3 percent. In the Q-learning based AODV protocol simulation, our Q-learning mechanism at each node self-configures the *active route timeout* and *hello interval* according to the Q-value. Due to the self-configured parameters, the nodes send RREQs more appropriately to account for failed routes, improving the route freshness and the link failure detection processes.

Table 1. Experimental Results: Standard AODV Protocol versus dynamic management

Items	AODV			Q-learning based Reconfiguration			Overall enhancement (%)
	#1 Mesh to Mesh	#2: Mesh to Ethernet	#3:Ethernet to Mesh	#1: Mesh to Mesh	#2:Mesh to Ethernet	#3:Ethernet to Mesh	
Route Discovery Time (s)	0.575856			0.41425			28.0
Route Errors (packets/s)	4.178218			1.178218			71.9
Control Overhead (packets/s)	42.17492			18.80528			55.3
Packet Jitter (sec)	0.247	0.0087	0.144	0.001	0.100	0.035	65.9
ETE delay (s)	1.550	0.576	0.856	0.061	0.406	1.216	43.7
SLA (%)	16.053	32.411	7.313	55.77	36.699	7.417	14.7
Packet Delivery Ratio (%)	64.774	84.2233	75.314	96.978	90.507	76.134	13.1

6.2 Protocol Performance

The performance of Q-learning based reconfiguration in AODV and of AODV protocol was evaluated in terms of packet delivery ratio, end-to-end delay and SLA compliance metrics. The packet delivery ratio is defined as the percentage of data packets successfully delivered to the intended destination. The end-to-end delay metric is defined as the average elapsed time between the generation and reception of data packets. From the results of Table 1, it is clear that the Q-learning based reconfiguration in AODV delivers at least 13 percent more packets than does the standard AODV. This result can be explained in terms of the network overhead due to the control messages generated by the AODV protocol. From the results presented in Table 1 it is clear that the AODV mechanism generates a greater number of control messages than does our AODV protocol with Q-learning based reconfiguration. This, in turn, translates into a higher probability of lost control messages in AODV due to collisions in the wireless medium. Consequently, routing paths are less reliable under AODV.

On average, our Q-learning based reconfiguration achieves an enhancement of end-to-end delay equivalent to 43.7 percent and SLA compliance equivalent to 14.7 percent. From these results it is clear that under the Q-learning based reconfigurable architecture, it is possible to achieve a lower end-to-end delay metric than under an AODV architecture. Reduced traffic in the wireless medium allows our AODV to realize a shorter queuing delay, resulting in shorter end-to-end delays. It is expected that our optimized AODV will outperform AODV in highly mobile networks because of its capability of reducing the control overhead and quickly reestablishing new routing paths, as well as efficiently using high-capacity links.

7 Conclusions and Future Research

In this paper, we have described our vision of a framework for autonomously reconfigured network systems with a cross-layer approach. The Q-learning based reconfiguration mechanism has been proposed to improve the performance of AODV. This is applicable to large heterogeneous networks, where the characteristics of the mobile nodes and application demands are different. We have also presented experimental results of our network system using OPNET. The performance results confirm that, in comparison to original AODV, our Q-learning based reconfiguration mechanism dramatically reduces the protocol overhead (55.3% reduction). It achieves a higher packet delivery ratio (13.1% enhancement) while incurring shorter queuing delays. More specifically, with the Q-learning based reconfiguration, it is possible to achieve shorter end-to-end delays (43.7% reduction) while reducing the incidence of lost data packets. Therefore, our autonomous reconfiguration mechanism successfully improves the scalability and adaptivity of the original AODV protocol in a heterogeneous network environment. Future research may include performance evaluation under diverse mobility patterns with other routing protocols (proactive, hierarchical and hybrid). The work in this paper highlights some interesting and potentially important areas for future work.

There is a fundamental tradeoff between proactive and reactive routing protocols, namely improved delay or reduced control overhead. A proactive routing protocol generates routing traffic independent of application traffic. Due to the higher routing overhead in proactive routing protocols (e.g. OLSR) we have chosen the reactive routing protocol, AODV, in our cross-layer approach to enhance the protocol performance. However, it is inevitable that certain static networks will have especially high QoS demands, which require the use of proactive routing. How to employ proactive routing while minimizing the network-layer overhead is of key interest. We plan to verify our scheme for other simulation scenarios, i.e. characterized by more heterogeneous networks (e.g. 3G, WiMAX and optical networks), various traffic models and mobility models.

The performed simulations were executed with a relatively low learning rate (α = 0.001). During the course of our experimentation, we saw that the Q-learning based reconfiguration was sensitive to learning rate reassignment. One solution to this problem is to use Bayesian exploration [19], to tune and optimize learning rate values. There is also a need to investigate optimization accuracy and the process of reward value assignment in the Q-value computation, in addition to the selecting of correct parameters for reconfiguration. To achieve desired QoS guarantees it is necessary to consider the change of user demands in the application layer over time. For consistent QoS support, MAC layer mechanisms can provide the essential per-link performance context, in the form of network congestion levels and achievable data rates.

While the work presented in this paper is but an initial step in the direction of a more general intelligent network management framework, the results are nonetheless promising. Eventually, successful simulation will lead to the creation of a hybrid network test-bed for real-world implementation. Cross-layer optimization, in the form of link-layer performance metrics and application-layer QoS demands, will enable a more comprehensive autonomous management framework to pursue issues such as dynamic clustering and inter-cluster service level agreements. Further refinement of our feedback mechanism is required to better optimize network-wide performance. This work is a building block upon which a general management framework can be built, one which is autonomously able to perform resource allocation and real-time network analysis.

Acknowledgments. This work was supported in part by the National Science Foundation (NSF) under Grant NeTS-0520333.

References

1. Wang, P., Wang, T.: Adaptive Routing for Sensor Networks using Reinforcement Learning. In: CIT 2006. Proceedings of the Sixth IEEE International Conference on Computer and Information Technology, vol. 00 (2006)
2. Demestichas, P., Dimitrakopoulos, G., Strassner, J., Didier, B.: Introducing reconfigurability and cognitive networks concepts in the wireless world. Vehicular Technology Magazine 1, 32–39 (2006)
3. Ryan, W.T., Daniel, H.F., Luiz, A.D., Allen, B.M.: Cognitive networks: adaptation and learning to achieve end-to-end performance objectives. Communications Magazine 44, 51–57 (2006)
4. C. W. a. P. Dayan: Machine Learning, pp. 279-292 (1992)
5. Conti, M., Maselli, G., Turi, G., Giordano, S.: "Cross-layering in mobile ad hoc network design. Computer 37, 48–51 (2004)
6. Hai, J., Weihua, Z., Xuemin, S.: Cross-layer design for resource allocation in 3G wireless networks and beyond. Communications Magazine 43, 120–126 (2005)
7. Borgia, E., Conti, M., Delmastro, F.: Mobileman: design, integration, and experimentation of cross-layer mobile multihop ad hoc networks. Communications Magazine 44, 80–85 (2006)

8. Srivastava, V., Motani, M.: Cross-layer design: a survey and the road ahead. Communications Magazine 43, 112–119 (2005)
9. Jack, L.B., Philip, F.C., Brian, K.H., William, T.K.: Key Challenges of Military Tactical Networking and the Elusive Promise of MANET Technology. Communications Magazine 44, 39–45 (2006)
10. Faccin, S.M., Wijting, C., Kenckt, J., Damle, A.: Mesh WLAN networks: concept and system design," Wireless Communications (see also IEEE Personal Communications). Wireless Communications 13, 10–17 (2006)
11. Forde, T.K., Doyle, L.E., O'Mahony, D.: Ad hoc innovation: distributed decision making in ad hoc networks. Communications Magazine 44, 131–137 (2006)
12. Basagni, S.: Distributed clustering for ad hoc networks, pp. 310-315 (1999)
13. T. C. a. P. Jacquet: Optimized Link State Routing Protocol (OLSR), IETF RFC 3626 (2003)
14. Maltz, D.J.a.D.: Dynamic Source Routing in Ad Hoc Wireless Networks. In: Imielinski, T., Korth, H. (eds.) Mobile Computing, pp. 153–179 (1996)
15. Perkins, E.B.-R.C., Das, S.: Ad hoc On-Demand Distance Vector (AODV) Routing. RFC 3561 (2003)
16. Haas, M.R.P.Z.J., Samar, P.: The Zone Routing Protocol (ZRP) for Ad Hoc Networks. IETF Internet-Draft (2002)
17. Boyan, J.A.L., L., M.: Packet Routing in Dynamically Changing Networks: A Reinforcement Learning Approach. In: dvance In Neural Information Processing System (1994)
18. http://www.opnet.com
19. Dearden, R., Friedman, N., Andre, D.: Model based Bayesian Exploration, UAI, pp. 150–159 (1999)
20. Koksal, C.E., Balakrishnan, H.: Quality-Aware Routing Metrics for Time-Varying Wireless Mesh Networks. IEEE Journal on Select Areas in Communications 24(11), 1984–1994 (2006)

Highspeed and Flexible Source-End DDoS Protection System Using IXP2400 Network Processor*

Djakhongir Siradjev, Qiao Ke, JeongKi Park, and Young-Tak Kim**

Dept. of Information & Communication Engineering, Graduate School, Yeungnam University, 214-1, Dae-Dong, Gyeongsan-Si, Gyeongbook, 712-749, Korea
m0446086@chunma.yu.ac.kr, keqiao817@hotmail.com, jk21p@ynu.ac.kr, ytkim@yu.ac.kr

Abstract. This paper proposes an architecture of source-end DDoS protection system on IXP2400 network processor, which monitors traffic from the source network and polices traffic at the source without affecting the traffic from other network. The proposed architecture includes usual IPv4 forwarder with additional modules for source filtering, packet classification and flow control, and uses modified non-parametric CUSUM algorithm. We analyze the major shortcomings of previous approaches, and present basic performance analysis. The proposed system can handle 65,000 aggregated flows, and can operate at OC-48 line rate.

Keywords: network security, DDoS, flow monitoring, classification, network processors.

1 Introduction

Nowadays, importance of the role of the Internet is rapidly increasing, due to continuous movement of major information flows into the Internet. As a result, the requirements to service availability and quality are rapidly increasing. Since the majority of Internet protocols have been developed without security consideration, they contain a lot of vulnerabilities that are exploited by various malicious users. Countering DDoS attacks is a very sophisticated task, complicated with similarity of malicious and legitimate traffic. Since DDoS attackers work in compliance with network protocols, it is impossible to distinguish malicious traffic. Another problem in detection DDoS attack is its distributed nature, because a single attacking host has limited network resource, which is usually less than server resources, while multiple attacking hosts can easily consume all of server resources. Rapid increase of data rates in access and core networks complicates the problem of dynamic flow monitoring, due to high number of per-packet operations. Scalability of flow monitoring system must also be supported, because of high number of flows going through network router.

* This research is supported by the MIC, under the ITRC support program supervised by the IITA (IITA-2006-(C1090-0603-0002)).

** Corresponding author.

D. Medhi et al. (Eds.): IPOM 2007, LNCS 4786, pp. 180–183, 2007.

There are three approaches used in DDoS protection [1]: (i) source-end, (ii) network core routers-based, and (iii) victim-end. Victim-end approaches can detect attacks, but can not protect the service, due to impossibility of distinguishing between legitimate and malicious connections. Network core routers based protection approach can identify and throttle attacking flows with average efficiency. Source-end approaches can efficiently detect attacking flows, and block them or limit their rate, in order to maintain service availability for users belonging to other networks. Recently, various algorithms [2,5,4,6] using network flow information have been proposed for usage in network core or source-end.

This paper proposes an architecture of source-end DDoS protection, based on modified *Cumulative Sum* (CUSUM) [2] algorithm, with dynamic flow monitoring and per-flow control for DDoS protection on IXP2400 Network Processor [3]. The main goals of designed architecture are high packet processing rate and scalable flow monitoring for DDoS protection, using features of IXP2400 Network Processor and considering its limitations. Also the system includes source filtering, to detect source IP address spoofing. The proposed design supports monitoring up to 65,536 aggregated flows, on the data rate of 2.4 Gbps. The design of packet processing pipeline and overall application is covered in details, and the algorithm of DDoS detection is explained.

The rest of this paper is organized as follows. In Section II, we describe the architecture of the proposed system, details of its functional modules, implementation details. We will cover the packet classifier implementation plan in details, as one of the main focuses of this paper. Section III analyzes the estimated performance of the proposed architecture, and discusses the shortcomings of other proposals. Finally, section IV concludes the paper.

2 Highspeed and Flexible Source-end DDoS Protection System Using IXP2400 Network Processor

The major requirements of DDoS protection are high speed, scalability, and high probability of correct attack detection. Our approach is source-end DDoS protection mechanism, which allows traffic monitoring of source subnetwork and police traffic at the source without affecting legitimate traffic. Being based on the usual IPv4 packet forwarder processing, our proposed DDoS protection system contains 3 additional modules: source filtering, classifier, and rate limiter. Fig. 1 shows component-based design of the proposed system. Protection from source IP address spoofing is required to prohibit usage of IP addresses belonging to other hosts to protect from indirect attacks, and reduce capabilities of attacker. Per-port source network information is maintained for checking source IP address of packets. Connection and aggregate flow classifier performs classification based on IP and transport layer header fields, gathers flow statistics, and decides if flow is malicious. Aggregate flow is determined by destination IP address, and connection is determined by usual 5 tuples of IP header. Classifier maintains legitimate traffic models and flow and connection statistics. Rate limiter polices malicious traffic. It receives packets detected as malicious, and limits

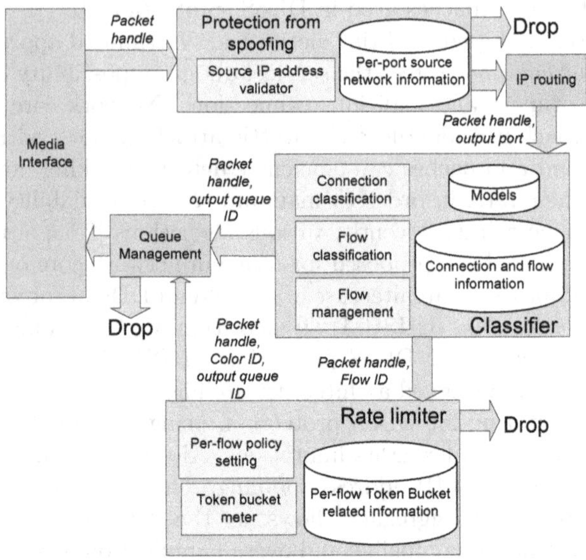

Fig. 1. Component-based design of DDoS protection system

their output rate, dropping packets that are exceeding the limitation. The application is implemented on IXDP2400 development platform. It has two IXP2400 [3] network processors. Due to the paper size limitations, detailed explanation of packet processing is omitted.

Our algorithm uses non-parametric CUSUM [2], but in contrast to other approaches it is based on sent and received packets ratio. Differing from D-WARD [4] it uses observations during predefined period of time to make the detection time independent from the previous results of observation. Increase of sent traffic over received is accumulated, and once it reaches predefined threshold, flow is detected as suspicious, and its rate limitation is applied to it.

3 Comparison and Evaluations

First, we discuss the shortcomings of currently known approaches for DDoS detection. Non-parametric CUSUM [2] for detecting is based on SYN-FIN and SYN-SYN/ACK pairs. The attacker can make their detection system fail by flooding a mixture of faked SYN, FIN, RST and SYN-ACK packets. [5] requires huge list of filtered IP source addresses database, thus increasing the complexity of algorithm. D-WARD [4] uses the ratio of sent and received TCP packets that is compared to predefined threshold. This may cause latency in attack detection when the flow has been existing for a long time. [6] proposes using non-parametric CUSUM based on SYN-ACK pairs. But authors do not consider the case when the number of malicious connections is less than the number of legitimate connections.

One of the major performance metrics for DDoS Protection system, is percentage of successfully detected malicious flows. Also the number of false positives is also important. In order to test application in the real conditions following testbed, containing Smartbits SMB6000 traffic generator, used for generating SYN flood, multiple TCP servers and legitimate clients, and interconnected using IXDP 2400 running the DDoS Protection system, is used. Servers under DDoS attack with and without protection system. Performance evaluations show that system is able to handle network traffic with 2.4 Gbps rate. With the rate limiting of suspicious flows to 5 fps, and with 1,000 legitimate and 1,000 malicious flows, system reduces fraction of rejected legitimate connections from 91% to 11%. System efficiency decreases when the number of malicious flows is much higher than number of legitimate flows. For better efficiency, adaptive rate limiting and threshold setting must be used.

4 Conclusion

In this paper we proposed DDoS protection system on Network Processor that uses dynamic flow monitoring and per-flow control, based on modified nonparametric CUSUM algorithm. The proposed system allows fine grain flow monitoring on 2.4 Gbps data rate, which allows it to successfully protect from DDoS attack, either at source-end or in the network core. The experiment result shows that in majority cases this system is able to process packet with the target line speed of 2.4 Gbps, supports maximum number of 65,000 flows and can greatly improve service availability to legitimate clients from other subnetworks. Our future works include using adaptive parameters for the algorithm, and more thorough comparison with other software and hardware-based approaches.

References

1. Mirkovic, J., Reiher, P.: A taxonomy of DDoS attacks and defense mechanisms. ACM SIGCOMM Computer Communications Review 34(2), 39–54 (2004)
2. Wang, H., Zhang, D., Shin, K.G.: Change-point monitoring for detection of DoS attacks. IEEE Transactions on Dependable and Secure Computing 1(4) (December 2004)
3. Intel IXP2400 Network Processor Hardware Reference Manual, Intel Corporation (October 2004)
4. Mirkovic, J., Reiher, P.: D-WARD: A source end defense against flooding denial-of-service attacks. IEEE Transactions on Dependable and Secure Computing 2(3), 216–232 (2005)
5. Peng, T., Leckie, C., Ramamohanarao, K.: Detecting distributed denial of service attacks by sharing distributed beliefs. In: Safavi-Naini, R., Seberry, J. (eds.) ACISP 2003. LNCS, vol. 2727, Springer, Heidelberg (2003)
6. Lim, B., Uddin, M.: Statistical-based SYN-flooding detection using programmable network processor. In: Proceedings of the Third International Conference on Information Technology and Applications, vol. 2, pp. 465–470 (2005)

Detecting Network Faults on Industrial Process Control IP Networks

Young J. Won[1], Mi-Jung Choi[1], Jang Jin Lee[2], Jun Hyub Lee[2], Hwa Won Hwang[3], and James Won-Ki Hong[1]

[1] Dept. of Computer Science and Engineering, POSTECH, Korea
{yjwon,mjchoi,jwkhong}@postech.ac.kr
[2] Electric & Control Maintenance Dept., POSCO, Korea
[3] Technical Research Laboratories, POSCO, Korea
{jangjin21,mujigae,hwawon}@posco.co.kr

Abstract. Industrial process control IP networks support communications between process control applications and devices. Communication faults in any stage of these control networks can cause delays or even shutdown of the entire manufacturing process. The current process of detecting and diagnosing communication faults is mostly manual and limited to post-reaction of user complaints which are followed by noticeable process malfunctioning. This paper identifies control network specific failures and their symptoms for early detection.

Keywords: Industrial Process Control Networks, IP Network Operation and Management, Fault Detection and Diagnosis, Traffic Monitoring.

1 Introduction

Process control networks support communications of devices in a controlling or manufacturing process. Existing process control network technologies (e.g., FOUNDATION Fieldbus, PROFIBUS, MODBUS, BACnet, LonWorks, etc.) use proprietary protocols and are not compatible with Ethernet and IP-based network technologies. However, newer versions of these technologies have adopted Ethernet (e.g., Industrial Ethernet) and IP for low cost, high scalability, and easy maintenance purposes. Despite of its wide deployment, we have a very little knowledge about managing fault-tolerant process control networks. This paper investigates process control networks of POSCO, the world's fourth largest iron and steel manufacturer, operating a number of plants world-wide. Their single operational site consists of more than 40 manufacturing plants, equally 40 process control networks where they are organized in a synchronous manner. A process control network is a combination of IP-based and non-IP based control technologies, so that the different monitoring and maintenance techniques should be used [1].

Network communication failures in any stage of process control can be fatal [5]. Since all machineries operate in a synchronous manner, a single communication failure to a device may delay or even force a shutdown of the entire plant process. Thus, it is much less fault-tolerant than the typical IP networks. For instance, an iron

D. Medhi et al. (Eds.): IPOM 2007, LNCS 4786, pp. 184–187, 2007.

and steel manufacturing plant involves a series of processes which are dependent on one another for the final product.

A typical process in a control network environment follow a hierarchical model where the controller at the top triggers actions in one or more controlled devices. The following are brief descriptions of the component elements and their roles.

- *Process Controller (PC)* – This is a part of the software and hardware package provided by the Programmable Logic Controller (PLC) vendors. It is the process control software running on a computer running UNIX or Windows that can remotely access PLCs. Custom-built or vendor provided server applications are placed in a PC to communicate with PLCs.
- *Programmable Logic Controller (PLC)* - It is a microprocessor computer for process control attached to a process control network. A complex sequence control of machinery (or low end controlled devices on factory assembly line) is handled by the custom-built software programs running in PLC.
- *Controlled Devices* – These machineries refer to sensors, actuators, motors, etc. They receive the command signals from PLCs via the embedded interface and perform various tasks.

We focus on identifying the problems at the first half of control networks (i.e., PC to PLC network) which is Ethernet and IP-based network. Although many IP network diagnosis tools (e.g., Sniffer [2], Wireshark [3]) are available, they often cannot detect the control network failure cases due to a significant difference of traffic nature as well as distinct communication failure characteristics. Yet no IP network diagnosis tools have been successfully handled control network specifics.

2 Real-World Control Network Fault Cases

We have analyzed the log history of process control network fault cases reported by the administrators at POSCO. Note that these cases were identified intuitively after the communication failures have occurred. The possible failures of process control networks are as follows:

- *Ethernet duplex mismatch* [6] - In the auto-configuration enabled environment, two end Ethernet devices may disagree about their duplex (half or full) settings after negotiation. Mismatch can increase the frame loss rate and add extra delays due to collision frames.
- *PLC programming bugs* - Poor PLC programming causes communication failures due to the lack of understanding and experience in socket programming. There have been a number of reports that might infer to the PLC programming bugs, such as unexpected packet occurrences (e.g., irregular keep-alive packets), disordered packet sequences, and unusual TCP window sizes.
- *Device driver bugs* - Device drivers in PLCs and controlled devices face an interoperability problem between various devices from different vendors. The robustness of device is also crucial for continuous operation.
- *Link corruptions* - These refer to physical damages to cables, e.g., cable cut, dust on fiber interface, etc. There are more chances of these incidents because process control networks are typically located in a hostile environment, such as a factory. We can only speculate the cable damage from measuring a few network metrics: invalid frame size, frame collision, CRC error, inconsistent throughput, etc.

- *Protocol unawareness* - If a particular device encounters unrecognizable protocols, it malfunctions and most likely causes a stoppage. It is important to prevent the unsupported protocol traffic, which are notified in the vendor's manual, from floating in the control networks.
- *Packet flooding* – The bandwidth is occupied by unwanted traffic, such as Internet worms.
- *Electrical noise* – Unstable transmission occurs due to signal interference. Network links, especially coaxial cables, and devices nearby high voltage machinery can be interfered with the noise.
- *Power outage* – Power supply of devices is malfunctioning. This problem is related to harsh conditions in process control networks, such as heat, moisture conditions, etc.
- *Misconfiguration* – Traffic is routed in non-optimal paths or even stays in a loop due to incorrect routing table entries. This results in packet delays and loss.
- *Damages to router/switch interface* – This is a typical hardware failure. It is somewhat difficult to detect because the whole connection to a certain area of network can be lost simultaneously.

Table 1. Control network specific measurement metrics and alarm conditions

Index	Network Metrics	Alarm Conditions
1	Collision frames	First appearance, or threshold-based
2	Jumbo (>= 1514 bytes) frames	First appearance
3	Runts (<= 64 bytes) frames	First appearance
4	CRC error frames	First appearance
5	IP/TCP checksum errors	First appearance
6	Fragment packets	Threshold-based
7	Retransmission packets	First appearance, or threshold-based
8	Packet inter-arrival time (ms)	Increase to the previous value
9	Throughput (bps)	Decrease, drop to 0, or pattern analysis over monitoring period
10	Packets per second (or packet burst)	Increase, decrease, drop to 0, or pattern analysis over monitoring period
11	Min/max/diff packet size (bytes)	Change in difference of max and min sizes over monitoring period
12	Min/max/diff TCP window size	Drop to 0, change in difference of max and min sizes over monitoring period
13	Out-of-order sequence packets	First appearance
14	Broadcast packets	Threshold-based
15	Unsupported protocol packets	Threshold-based

3 Early Symptoms of Communication Failures

It is important for network administrators to recognize early patterns of process network communication failures. As illustrated in Table I, we have selected several IP network metrics and identified the conditions accordingly that best reflect any irregularity of communication in process control networks [4]. Metrics themselves are not unique, but they have not been properly analyzed in many IP network diagnosis tools. A new set of monitoring categories and conditions is necessary because even a popular tool, like Sniffer, could not fully detect the control network specific failures but generate false network alerts. The selected metrics can be measured using passive monitoring techniques which do not interfere with network operations.

The metrics in index 8-12 have distinct conditions compared to those of IP data networks. In 24 hours of rolling out process, the communications between PC and PLC tend to be continuous with a steady inter-packet generation time. They also yield very low bandwidth communications with a fixed packet size. The listed conditions here indicate that any sudden changes (up or down) of value are considered abnormal in process control networks.

There is hardly a case where a single metric infers the cause of failure. They occur in a chain of reactions; for example, if the error frames or the out-of-order packets arrive, there is a high chance of seeing retransmission trials soon after. Consequently, the whole sequence of transmissions might have been triggered from the cable cut or software bugs.

4 Concluding Remarks

The area of monitoring and analyzing process control IP networks has been blinded in the research community thus far. Process control networks are much more vulnerable system in case of network outage because the consequences can be very costly. An early diagnosis of potential communication failures is crucial to maintain fault-tolerant network operating environment. This paper identified process control network specific failure types and alarm conditions (metrics) for possible communication failures. For future work, we plan to design and implement a remote diagnosis system that detects early symptoms of control IP network communication failures.

Acknowledgements. This research was supported in part by the MIC (Ministry of Information and Communication), Korea, under the ITRC (Information Technology Research Center) support program supervised by the IITA "(Institute of Information Technology Assessment) (IITA-2006-C1090-0603-0045)."

References

1. Lian, F.-L., Moyne, J.R., Tilbury, D.M.: Performance Evaluation of control networks: Ethernet, ControlNet, and DeviceNet. IEEE Control System Magazine 117(6), 641–647 (2001)
2. Network General Sniffer, http://www.networkgeneral.com/
3. Wireshark, http://www.wireshark.org/
4. Soucek, S., Sauter, T., Rauscher, T., Scheme, A.: A Scheme to Determine Qos Requirements for Control Network Data over IP. IEEE Industrial Electronics IECON 1, 153–158 (2001)
5. Plönnigs, J., Neugebauer, M., Kabitzsch, K.: Fault Analysis of Control Networks Designs. In: ETFA 2005. Proc. of the 10th IEEE International Conference on Emerging Technologies and Factory Automation, Catania, Italy (September 2005)
6. Detecting Duplex Mismatch on Ethernet, http://www.pam2005.org/PDF/34310138.pdf/

OSPF-AN: An Intra Domain Routing Protocol for Ambient Networks

Auristela Silva[1], Tarciana Silva[1], Luciana Oliveira[1], Reinaldo Gomes[1], Igor Cananéa[1], Djamel Sadok[1], and Martin Johnsson[2]

[1] Federal University of Pernambuco, Brazil,
[2] Ericsson Research, Sweden
{auristela,tarciana,lpo,reinaldo,icc,jamel}@gprt.ufpe.br, martin.johnsson@ericsson.com

Abstract. This short paper presents OSPF-AN protocol, an extension for OSPF protocol aiming for working in Ambient Networks. OSPF-AN adds services information in its database and messages, and as the original OSPF, will be applied in structured networks. This protocol could be applied together with other routing protocols in a scenario of dynamic networks. OMNet++ simulator was used to conduct simulation experiments, thus validating OSPF-AN.

Keywords: Routing Protocol, Ambient Networks, Services.

1 Introduction

The Ambient Networks project [1] represents a new networking concept, which defines a set of features, architectures and solutions, aiming at offering interoperability among heterogeneous networks belonging to different operators, providers or technology domains. The resultant network must seem homogeneous to the potential users of the network services. Many research groups have been working to define and validate coherent solutions for ambient networking based on a range of different features, such as policy, mobile management, security, self-organization, self-configuration, routing algorithms and others.

Considering different technologies, we could classify networks into many groups, such as structured, ad-hoc, mobile and short ranged networks. However, all of these networks need to cooperate in order to satisfy service requests of users. One way for this cooperation to occur is to re-examine routing across such a wide range of networks. In this short paper we present an extension of the OSPFv2 [2], a widely used routing protocol designed for wired networks, which we called OSPF-AN (OSPF for Ambient Networks).

OSPF is a link-state intra-domain routing protocol used in IP networks, which calls for the sending of Link State Advertisements (LSAs) to all other routers within the same hierarchical area. Information on attached interfaces, metrics used, and other variables are included in OSPF LSAs. In this context, service information was included into LSAs, and this set of LSAs builds the OSPF-AN database. Similarly to OSPF, the OSPF-AN accumulates link-state information, and uses the Dijkstra's

D. Medhi et al. (Eds.): IPOM 2007, LNCS 4786, pp. 188–191, 2007.

Shortest Path First algorithm to calculate the shortest path to each node. Additionally, OSPF-AN is able to interact with legacy systems using OSPF.

2 OSPF-AN: Overview

OSPF-AN introduces a proposal for extending the OSPF protocol in order to allow it to be used like an Internal Gateway Protocol in Ambient Networks, introducing service information in its routing. Consequently, OSPF-AN associates services to routes, not nodes or physical addresses. In our work, we classify services into two classes: user services (e.g. VoIP, VoD) and support services (e.g. QoS, security).

In this new routing approach, each router knows about the services belonging to its network and after the convergence time, each router has knowledge about service information from the other networks connected to it (see Figure 1). From this new database, considering links states and services, each router builds a routing table, where the service is the key, by calculating a shortest-path tree to reach that specific service, with the root of the tree being the calculating router itself.

When a router receives a request related to a given service, it searches in its routing table for this information, and discovers if the service is in its network or if the requested service belongs to another network. In case a router does not find the service in its routing table, it sends the request to external routers (called AS Boundary Router in OSPF), which have communication with another domains.

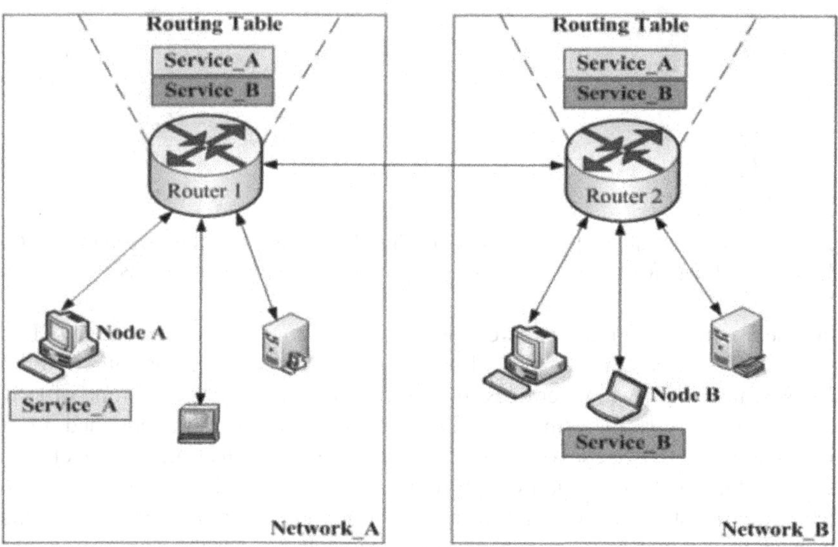

Fig. 1. Routers using OSPF-AN Protocol

3 Simulation and Initial Results

For simulation of the OSPF-AN we use the discrete event simulation environment, OMNeT++[3], which is free of charge for teaching and research use. We considered different topologies and different hierarchy structures of networks.

In the simulations we used two different scenarios: (1) Géant[4] Dark Fibre network with sixteen routers and (2) Géant network with thirty-four routes. The number of services that each router owns, in our experiments, ranges from one to eight. Since the network is composed by many routers, then each network might own several services.

In first simulation, we measure number of exchanged LSAs and convergence time only using OSPF-AN. In this case, both scenarios were evaluated utilizing one service in each router and after, several services in each router (we had router with one, two, three until eight services). Considering the obtained results, it is possible seeing that the number of exchanged LSAs remains the same independent on the number of services announced (Figure 2), showing that the mechanism of transmit services information using LSAs worked properly.

On the other hand, the convergence time (Figure 3) increased as long as the number of services increased. Utilizing only one service in each router, the convergence time remained around 5s. With several services per router, the convergence time was 40s. This difference cannot be ignored, but we should analyse that the increasing of announced information was proportional to the increasing in time, so we can consider it an acceptable convergence time.

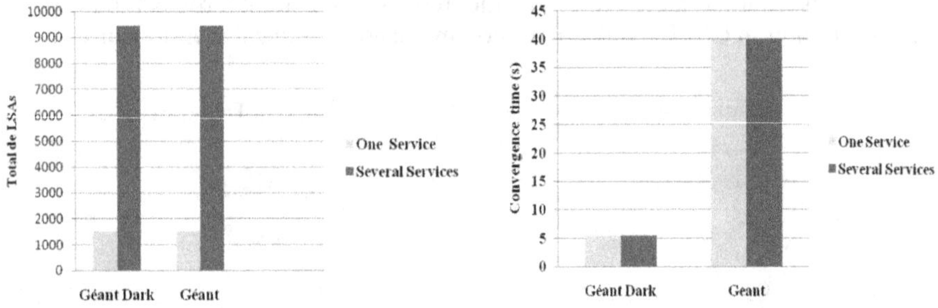

Fig. 2. Total of exchanged LSAs in Géant Dark Fiber and Géant networks

Fig. 3. Convergence time (seconds) in Géant Dark Fiber and Géant networks

In the second simulation, we measure the number of exchanged LSAs, convergence time, bandwidth, total of bytes in same scenarios, comparing OSPF and OSPF-AN, utilizing several services in each router. The convergence time of OSPF-AN was around 40s in terms of simulation time (same value obtained in first experiment, when utilized several services in each router). This time was much bigger than in OSPF, which converged in approximately 5s. This value was the same found in first experiment, when we used only one service in each router. So we concluded that when each router has one service, both protocols present the same convergence time.

This difference between both protocols can be explained since for each router (in each interface or IP address) we associated services, which did not exist previously in OSPF. These services have also to be stored in routing tables, what will create routing tables with more information and then expanding the time necessary to fill all the existent network information (services and IP addresses).

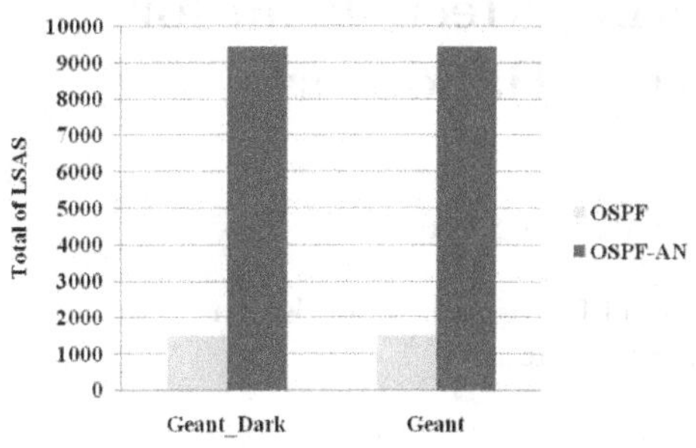

Fig. 4. Total of exchanged LSAs in Géant Dark Fiber and Géant networks

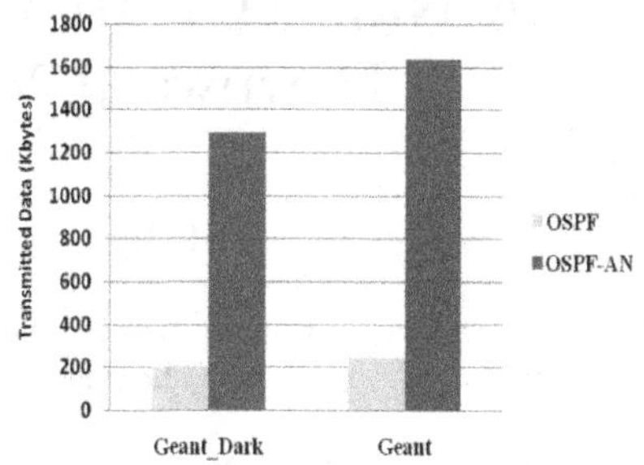

Fig. 5. Transmitted Data (Kbytes) in Géant Dark Fiber and Géant networks

We also collected information related with messages and traffic necessary to exchange information by the protocols in our simulations. In Figure 4 we have the amount of LSAs exchanged. It is possible verifying that the number of exchanged LSAs is the same, independent of the presence of services information, when observing both protocols being used in the same scenario.

In Figure 5 we analyzed the total transmitted data in Kbytes. The transmission of service by OSPF-AN increases the information exchanged between routers. The transmitted bytes increased 21% into *Géant Dark Fibre* and 26% in *Géant Network.*

4 Conclusion

In future works, we intend to enhance our simulations of the OSPF-AN protocol, doing more comparisons with traditional OSPF, and evaluating, in a real network, features such as *convergence time*, *cost of the protocol*, specifically consumed *link bandwidth*, and also how the OSPF_AN protocol will perform and scale in the Internet. Different scenarios will be used, including the integration with an inter domain protocol for Ambient Networks.

References

1. WWI-AN Ambient Networks Project WWW Server, http://www.ambient-networks.org
2. RFC 2328, J. Moy. OSPF Version 2. Request For Comments 2328, Internet Engineering Task Force, April (1998)
3. OMNet++ Community Site, www.omnetpp.org
4. Géant Project, http://www.geant.net/

A New TCP Reordering Mechanism for Downward Vertical Handover*

HoJin Kim and SuKyoung Lee

Dept. of Computer Science, Yonsei University, Seoul, Korea
hojin@winet.yonsei.ac.kr

Abstract. In integrated WLAN and cellular networks, a downward vertical handover (DVHO) causes an abrupt change so that packets which are transmitted through both networks can be out of ordered. Reordering problem triggers unnecessary fast retransmission causing throughput degradation. Thus, we propose a TCP reordering mechanism for DVHO, that suppresses unnecessary retransmissions due to the duplicate acknowledgements (dupacks). The simulation shows that the proposed TCP achieves better performance compared with nodupack with SACK.

Keywords: Downward vertical handover, TCP, Reordering.

1 Introduction

The increasing demand for high capacity wireless Internet access has led to the integration of 3G/4G cellular networks with wireless local area networks (WLANs). After a downward VHO (DVHO) from the cellular network to the WLAN, some packets transmitted over the WLAN are likely to overtake the acknowledgements (ACKs) for the preceding packets still in transit over the cellular network due to the shorter Round Trip Time (RTT) in the WLAN. Thus, TCP sends duplicate acknowledgements (dupacks) causing to falsely trigger fast retransmission, which decreases the congestion window size (*cwnd*) and the slow start phase threshold (*ssthresh*). The authors of [1] and [2] propose to use Duplicate Selective Acknowledgement (DSACK) [3] to improve TCP performance by adjusting the threshold of dupacks (*dupthresh*) and inserting a small delay before retransmitting the lost packet. However, the *dupthresh* based on the Exponentially Weighted Moving Average (EWMA) is not suitable because bandwidth is abruptly changed after DVHO. In [4], transmission of dupacks is suppressed during a VHO up to *dupthresh*. But it is not mentioned how to set an appropriate value of *dupthresh*. We propose a new TCP reordering mechanism for DVHO, that avoids activating false fast retransmission due to the spurious dupacks by making TCP receiver send normal ACKs for the out-of-order packets received from the WLAN after a DVHO, rather than send dupacks because the in-flight packets would be received soon over the cellular network. We show

* This work was supported by grant No.R01-2006-000-10614-0 from the Basic Research Program of the Korea Science & Engineering Foundation.

D. Medhi et al. (Eds.): IPOM 2007, LNCS 4786, pp. 192–195, 2007.

via simulations that the proposed TCP improves throughput performance after DVHO over existing scheme.

2 TCP Reordering Mechanism for DVHO

Packets sent after a DVHO are very likely to overtake the packets still in flight over the cellular network because of the shortened RTT in the WLAN. As a response to out-of-order packets, dupacks are sent, carrying the same ACK number as the previous ACK, which triggers the fast recovery algorithm until all the in-flight packets arrive, wasting the bandwidth in the WLAN because it reduces the *cwnd* in half. For the sender to avoid the false fast retransmission after DVHO, we propose to send normal ACKs for the out-of-order packets after DVHO, that suppresses dupacks until in-flight packets arrive. The number of in-flight packets can be obtained by the difference between the sequence number of the last packet received from the cellular network and that of the first packet received from the WLAN. At that time, the information is delivered in the TCP header for the receiver after DVHO to inform the sender the number of in-flight packets in the cellular network.

In case that in-flight packets are lost, the TCP sender maintains the timer in the cellular network separately with that in the WLAN. If in-flight packets has not arrived yet at the TCP receiver and the elapsed time since the in-flight packet exceeds the timeout value in the cellular network, then the packet is regarded lost. In other words, unless the TCP sender maintains the timer in cellular network, in-flight packets can be retransmitted over WLAN even though in-flight packets are being arrived over cellular network because the retransmit timer is reset shortly in WLAN by shortened RTT. Therefore, dual timer in both networks can prevent unnecessary retransmission as well as resolve loss problem of in-flight packets. When the in-flight packet loss occurs, TCP sender retransmits the remaining in-flight packets over WLAN and the timer in WLAN is reset.

Let d_i and d_j (i and j indicate the sequence number) be the last packet and the first packet which the Mobile Node (MN) receives over the cellular network and WLAN, respectively. TCP-D refrains dupacks after DVHO and moves the *cwnd* although out-of-order packets, d_k $(k > j)$ arrive at the TCP sender. Instead of sending dupacks as a response to the out-of-order packets, TCP-D waits for the in-flight packets, $d_{i+1}, d_{i+2}, ..., d_{j-1}$. The operations of TCP-D sender and receiver are as follows:

<u>TCP receiver</u>

1. If an MN becomes aware that a DVHO is about to occur upon the receipt of L2 (link layer) trigger (i.e. received signal strength or beacon signal), it sends a DVHO message to the TCP sender with handover notification.
2. Even if $j > (i + 1)$, the d_j is regarded out-of-ordered and the TCP receiver sends an ACK for the packet d_j including N where N is $j - i - 1$.

3. When a packet received by the node is one of the in-flight packets (i.e., $d_{i+1}, d_{i+2}, ..., d_{j-1}$), N decreases by 1. Otherwise, the MN acknowledges the received packet, d_k $(k > j)$ without changing the value of N.

TCP sender

1. When the TCP sender receives the ACK for d_j, it can send the next packet and start its timer where the retransmission timeout (RTO) is updated based on the sample RTT measured for the packet, d_j over the WLAN.
2. If an ACK for d_k $(k > j)$ with $N > 0$ is received, the TCP sender will send the next packet, d_{k+1}. If in-flight packets d_n $(i + 1 < n < j)$ are regarded lost, TCP sender retransmits the remaining in-flight packets including the first lost packet (i.e., $d_{n+1}, d_{n+2}, ..., d_{j-1}$) over WLAN. The timer in WLAN is reset for the retransmit packet. In addition, to avoid degrading link utilization in WLAN, *cwnd* and *ssthresh* are maintained as previous values instead of returning back to the normal TCP.
3. When an ACK with $N = 0$ is received, TCP-D goes to the normal TCP operation because all in-flight packets from d_{i+1} to d_{j-1} have been received.

Even though reordering scheme resolves the throughput degradation problem well, inbursty packets in TCP receiver is probably sent. The TCP sender moves *cwnd* up to the number of overtaking ACKs over WLAN when the transmission of in-flight packets over cellular networks is completed because TCP receiver already sends normal ACKs for the packets sent over WLAN. To prevent TCP sender from injecting bursty packets and efficiently utilize the increased bandwidth of the WLAN, we propose two schemes in sender-side. Once the DVHO is completed, TCP sender enters the Slow Start (SS) phase with its $cwnd = 1$ and *ssthresh* readjusted based on the Bandwidth Delay Product (BDP) of the WLAN estimated on the link layer. Further, we use *dual queue* to prevent bursty injection. The TCP sender transmits packets through single queue (Q1) and enables another queue (Q2) on the receipt of handoff notification. Then, TCP sender transmits packets through Q2 and moves *cwnd* of Q2 according to the ACKs. When the ACKs for packets buffered in Q1 arrives, TCP sender informs Q1 of disable notification. When a loss indication of the packets buffered in Q1 occurs, the TCP-D retransmits the lost packet, but basic congestion control does not work at that time. A loss of in-flight packets does not affect current sending rate because the loss of in-flight packets is caused by handoff latency. On the other hand, When a loss indication of the packets buffered in Q2 occurs, TCP-D operates according to basic TCP congestion control. Consequently, using proposed two schemes, TCP-D reaches to stable state rapidly, preventing spurious fast retransmission.

3 Performance Evaluation

In this section, we provide simulation results to show the improved throughput performance of the proposed TCP mechanism using *ns*-2.28 with wireless extensions. The available bandwidth in WLAN is 2 Mbps with end-to-end RTT of

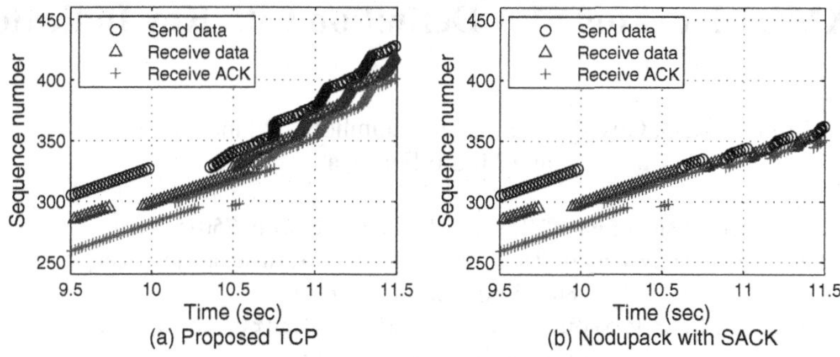

Fig. 1. Improvement TCP Performance

100 msec while cellular networks bandwidth is 384 Kbps with end-to-end RTT of 300 msec, refereing to UMTS. A DVHO occurs at 10 sec and TCP-D initiates SS phase after DVHO so that it estimates available bandwidth after DVHO rapidly. Fig. 1 shows a plot of the sequence number sent and the ACK number received at the TCP sender and the sequence number received at the TCP receiver respectively for (a) the TCP-D and (b) the nodupack with SACK [4] in 384 Kbps → 2 Mbps. In TCP-D, in-flight packets are being received while out-of order packets in new link are sent at 10.4 sec. However, normal ACKs are sent instead of dupacks so that *cwnd* is increased according to the number of ACKs sent over the WLAN. On the other hand, nodupack with SACK cannot increase *cwnd* because no normal ACKs are sent while out-of-order packets arrive even though dupacks are not sent. Then, ACKs for the packets in receiver buffer are sent at once when all in-flight packets arrive. Consequently, it is observed that the throughput under the TCP-D is higher than nodupack with SACK.

4 Conclusion

In integrated WLAN and cellular networks, an MN suffers from packet reordering after a DVHO due to abrupt change in link bandwidth. In this paper, we propose to new TCP reordering mechanism which resolves the reordering problem to prevent throughput degradation due to spurious dupack after a DVHO.

References

1. Blanton, E., Allman, M.: On Making TCP More Robust to Packet Reordering. ACM Computer Communication Review 32(1) (January 2002)
2. Ma, C., Leung, K.: Improving TCP Robustness under Reordering Network Environment. In: IEEE GLOBECOM 2004 (2004)
3. Floyd, S., Mahdavi, J., Mathis, M., Podolsky, M.: An extension to the selective acknowledgement (SACK) option for TCP, RFC2883 (July 2000)
4. Hansman, W., Frank, M.: On Things to happen during a TCP Handover. IEEE LCN 2003 (2003)

An XML Model for SLA Definition with Key Indicators

Emir Toktar[1], Guy Pujolle[1], Edgard Jamhour[2], Manoel C. Penna[2], and Mauro Fonseca[2]

[1] University of Paris VI, 8, rue du Capitaine Scott, 75015, Paris
[2] Pontifical Catholic University of Paraná, PUCPR, PPGIA, Imaculada Conceição St. 1155, 80215-901, Curitiba, Brasil
emir.toktar@computer.org, emir.toktar@etu.upmc.fr, Guy.Pujolle@lip6.fr, {jamhour,penna,mauro.fonseca}@ppgia.pucpr.br

Abstract. This work proposes a XML-based model for the specification of service level agreements (SLA). The model has XML elements to define a semantic to represent key performance indicators (KPI) and key quality indicators (KQI) and the relationship between them. *Upper* and *lower* thresholds are associated to the indicators in order to indicate warnings or errors conditions. The relationship between the indicators is expressed by reusable functions which are evoked by the XML-based model. An example of reusable function for calculating the KQI service availability based on KPI indicators is also presented in this paper.

Keywords: SLA, KPI, KQI, XML.

1 Introduction

Important research efforts on Quality of Service (QoS) management oriented to IP infrastructure management have been developed in the last years. QoS management is the way to assure the quality of delivered services by monitoring quality indicators that reflect a customer-service provider common understanding on what the service quality is. Service Level Agreement (SLA) provides the basis for QoS management according this perspective, establishing a two-way accountability for service that is negotiated and mutually agreed upon by customer and service provider.

However, when expressing QoS, service level indicators must reflect what is delivered by a particular network making necessary a relationship between service level indicators and network performance parameters. The mapping from SLA needs to network configuration is usually done manually and is a difficult and error prone task. Automatic translation of service level indicators into network configuration parameters is still an open problem, as well as relating network performance parameters with service level indicators. This mapping is not straightforward because the first evaluates quality of flow from the network perspective while the second reflects a negotiated view of service quality. An SLA information model that takes into account this issue is needed together with the algorithms that perform the necessary mapping.

D. Medhi et al. (Eds.): IPOM 2007, LNCS 4786, pp. 196–199, 2007.

The main concern of this study is related with mapping from network performance parameters to SLA high-level indicators. The objective is to present a XML based SLA model that uses Key Quality Indicator (KQI) and Key Performance Indicator (KPI) concepts for specifying service level indicators and network performance parameters, respectively, which includes the necessary elements to specify the translation from KQI into KPI.

2 Related Works

Several industrial and research works have been developed around QoS subject by using the SLA concept for QoS management, including communication networks and information technology. Proposal from many international projects are summarized on the EGEE publication [1], and relevant work can be found in the proposals from the TMForum [2] and 3GPP [3].

The AQUILA [4] project defined SLS (Service Level Specification) templates to standardize the requests of QoS between the customer and service provider, for the support of QoS in IP networks. The proposed approach groups IP applications by similar QoS behavior and requirements. This idea is similar to our model that defines templates for key performance indicators. The WSLA [5] is a XML-based Web Services SLA Language (WSLA) that can be used to model web SLAs specification and to be monitored through Web Services technology. WSLA is defined as an XML Schema and covers the definition of the involved parties, the services guarantees and the service description. The CADENUS [6] project considers a configuration and provisioning integrated solution for end users QoS services. The Service/SLA Model uses specialized services classes and SLS models to compose a SLA.

The above studies are just examples of the many different approaches that have been considered around the theme, generating massive concept diffusion, nevertheless with an elevated divergence in the definitions and models adopted. In this work, the SLA concept is used the definitions contained in the SLA Management Handbook [8]. As the focus is on the service relationship involving client/provider, an important question arises, that is, how to select adequate metrics for this relation. In this context the concept of *key indicator* is extended to facilitate the mapping between high-level service parameters through specific technology parameters by two new indicators, *key quality indicator* (KQI) and *key performance indicator* (KPI), which were introduced by the TeleManagement Forum [7].

KQI is a high-level indicator that captures the customer-service provider common understanding of QoS. On the other hand, KPI is directly related to performance characteristics of service elements. In a general way, a KQI is defined from a set of KPIs, that is, to calculate a KQI value, KPIs values are combined by empirical or theoretical functions. In this study, we specify KPIs to KQI translation this translation by using references to KPI and KQI in XML schemes.

Moreover, each KQI or KPI is associated with a set of thresholds, including lower and upper warning and lower and upper error. Theses thresholds are integrant part of our SLA model. QoS is agreed by defining KQIs and KPI, their corresponding thresholds, and the mapping.

3 KISLA Model

The model proposed in this work is called KISLA (Key-Indicator SLA), which is implemented in a XML-based language. The xml elements are used for describing the entities and relationships related to SLAs. The KPI and KQI definitions and evaluation are preformed through calls to reusable functions embedded in the markup language. The functions itself are not described in XML, but they are just evoked by the KISLAML interpreter, approach is inspired by the XACML [8]. In this section we illustrate a piece of KISLA model.

<SLAContractType> is the main class and aggregates three main classes: <Parties>, <Services> and <Responsabilities>. <Parties> describes customer and provider involved in an SLA. It aggregates other classes used to store information such as name, phone number, address, email and other related data, expressed using the classes proposed by the CIM version 2.9. <Services> represents the information about the offered service. It defines the service topology, the validity period for the SLA, and the service level. Service topology is formed by service elements and their corresponding KPIs, for example, delay, jitter, packet loss and average bit rate. <Responsabilities> represents the conditions that must be respected by the provider with respect to the offered service. These conditions are expressed in terms of KQIs, which are used for defining the terms under which the offered service will be monitored and evaluated. KQIs are expressed, preferentially, but not exclusively, as a function of KPIs. Availability is KQI example that can be defined in terms of delay, packet loss and average bit rate. Besides conditions, it also defines the penalties to be applied when the expressed conditions are offended.

<Responsabilities> also aggregates three classes <KeyQualityIndicator>, <Schedule> and <Duties>. The first defines how a KQI is calculated (see explanation below) in terms of the associated KPIs, included in <KeyPerformanceIndicatorSet>. The second defines the periodicity for KQI evaluation (e.g., daily, weekly, etc.) and the last defines the penalties, as described by <Penalty>, to be applied to the provider when the <Responsabilities> are violated. The penalties can be calculated as a function of the fees paid by the customer, as defined by <Cost>.

In order to simplify the process of defining KPI sets, the KISLA model proposes the use of XML schemas for automatic validating the KPI definitions. The *KPITemplateSet* is associated to a XML schema that can be selected from a library, when a service is instantiated. *KPITemplateSet* defines a set of reusable KPIs that can be used for describing similar services. For example, the family CODEC G defined by the ITU-T, specifies a set of CODEC specifications with similar parameter definitions. Therefore, CODECG is an example of KPITemplateSet and CODEC G729 is an example of KPITemplateSet instantiation. The structure of a KPITemplateSet is used to specify the parameters of a service element. For example, the ITU-T CODEC G series have three KPIs, <Packet_Loss>, <Jitter> and <Delay>, which are used to assure the "audio" quality. Each KPI is associated to a <SLOType> that defines the *thresholds* associated with it. The acceptable metric units for each threshold are also defined in the schema.

As an example, the template discussed above could be used for defining the KPI set for audio applications according to the ITU-T standard CODEC G.729. The standard specifies delay limited to ***150 ms***, jitter limited to ***50 ms*** and packet loss limited to ***0.5%*** of the transmitted packets. These KPI values specify the conditions the provider must assure for the customer. In order to satisfy these conditions, the provider must translate these parameters into SLS configurations, and apply them to the network devices that are affected by the ServiceAccessPoint associated to the service. For example, the SLS configuration for the G.729 service correspond to a "token bucket" with the following parameters specified on RFC2212 (rate=2000, bucket=80, peak=4000). The translation process for calculating the SLSs form the SLA definitions is out of the scope of this paper.

After KPIs definition, the related KQI can be calculated by using the evaluation function specified in <KQIFunction>, which is a pointer to a library function called by KISLA interpreter.

4 Conclusion

Defining a model for represent SLA information is an important step for creating management tools for automating the process of provisioning and monitoring *DiffServ* networks. The work in this paper presents a proposal towards a unified model for representing SLAs called KISLA that adopts the concept of representing SLAs by using the KQI and KPI indicators. This approach shows a great flexibility for describing a large number of services negotiated in a *DiffServ* domain. It provides means for defining a common nomenclature for negotiations between customers and providers.

The KISLA markup language was inspired by the experience acquired by building policy-based management systems based on both, purely object-oriented approaches, such as CIM, and hybrid-approaches, capable of evocating external reusable functions, such as XACML. We have seen advantages in the hybrid approach, because it permits to create policies where most semantic definition is defined by the policy language, instead of being implicitly defined by the policy interpretation algorithm. This approach was particularly useful for adapting the model to the KPI and KQI definitions.

References

[1] EGEE, Enabling Grids for E-science in Europe, URL http://www.eu-egee.org
[2] TeleManagement Forum, URL http://www.tmforum.org
[3] 3GPP. 3rd Generation Partnership Project (3GPP), URL: http://www.3gpp.org
[4] Salsano, S., et al.: Definition and usage of SLSs in the AQUILA consortium, draft-salsano-aquila-sls-00.txt (November 2000), URL: http://www.ist-aquila.org/
[5] Ludwig, H., Keller, A., Dan, A., Franck, R., King, R.P.: Web Service Level Agreement (WSLA) Language Specification. IBM Corporation (July 2002)
[6] CADENUS project, References at URL: http://wwwcadenus.fokus.fraunhofer.de
[7] TeleManagement Forum, S.L.A.: Management Handbook, Enterprise Perspective, G045. The Open Group. 4 (October 2004)
[8] XACML - OASIS. eXtensible Access Control Markup Language (XACML) Version 2.0. OASIS (February 2005), URL: http://www.oasis-open.org/committees/xacml/

Author Index

Lecture Notes in Computer Science

Sublibrary 5: Computer Communication Networks and Telecommunications

For information about Vols. 1–4503
please contact your bookseller or Springer

Vol. 4793: G. Kortuem, J. Finney, R. Lea, V. Sundramoorthy (Eds.), Smart Sensing and Context. X, 301 pages. 2007.

Vol. 4787: D. Krishnaswamy, T. Pfeifer, D. Raz (Eds.), Real-time Mobile Multimedia Services. XII, 197 pages. 2007.

Vol. 4786: D. Medhi, J.M. Nogueira, T. Pfeifer, S.F. Wu (Eds.), IP Operations and Management. XII, 201 pages. 2007.

Vol. 4785: A. Clemm, L.Z. Granville, R. Stadler (Eds.), Managing Virtualization of Networks and Services. XIII, 269 pages. 2007.

Vol. 4773: S. Ata, C.S. Hong (Eds.), Managing Next Generation Networks and Services. XIX, 619 pages. 2007.

Vol. 4745: E. Gaudin, E. Najm, R. Reed (Eds.), SDL 2007: Design for Dependable Systems. XII, 289 pages. 2007.

Vol. 4725: D. Hutchison, R.H. Katz (Eds.), Self-Organizing Systems. XI, 295 pages. 2007.

Vol. 4712: Y. Koucheryavy, J. Harju, A. Sayenko (Eds.), Next Generation Teletraffic and Wired/Wireless Advanced Networking. XV, 482 pages. 2007.

Vol. 4686: E. Kranakis, J. Opatrny (Eds.), Ad-Hoc, Mobile, and Wireless Networks. X, 285 pages. 2007.

Vol. 4685: D.J. Veit, J. Altmann (Eds.), Grid Economics and Business Models. XII, 201 pages. 2007.

Vol. 4581: A. Petrenko, M. Veanes, J. Tretmans, W. Grieskamp (Eds.), Testing of Software and Communicating Systems. XII, 379 pages. 2007.

Vol. 4572: F. Stajano, C. Meadows, S. Capkun, T. Moore (Eds.), Security and Privacy in Ad-hoc and Sensor Networks. X, 247 pages. 2007.

Vol. 4549: J. Aspnes, C. Scheideler, A. Arora, S. Madden (Eds.), Distributed Computing in Sensor Systems. XIII, 417 pages. 2007.

Vol. 4543: A.K. Bandara, M. Burgess (Eds.), Inter-Domain Management. XII, 237 pages. 2007.

Vol. 4534: I. Tomkos, F. Neri, J. Solé Pareta, X. Masip Bruin, S. Sánchez Lopez (Eds.), Optical Network Design and Modeling. XI, 460 pages. 2007.

Vol. 4517: F. Boavida, E. Monteiro, S. Mascolo, Y. Koucheryavy (Eds.), Wired/Wireless Internet Communications. XIV, 382 pages. 2007.

Vol. 4516: L.G. Mason, T. Drwiega, J. Yan (Eds.), Managing Traffic Performance in Converged Networks. XXIII, 1191 pages. 2007.

Vol. 4503: E. Airoldi, D.M. Blei, S.E. Fienberg, A. Goldenberg, E.P. Xing, A.X. Zheng (Eds.), Statistical Network Analysis: Models, Issues, and New Directions. VIII, 197 pages. 2007.

Vol. 4479: I.F. Akyildiz, R. Sivakumar, E. Ekici, J.C.d. Oliveira, J. McNair (Eds.), NETWORKING 2007. Ad Hoc and Sensor Networks, Wireless Networks, Next Generation Internet. XXVII, 1252 pages. 2007.

Vol. 4465: T. Chahed, B. Tuffin (Eds.), Network Control and Optimization. XIII, 305 pages. 2007.

Vol. 4427: S. Uhlig, K. Papagiannaki, O. Bonaventure (Eds.), Passive and Active Network Measurement. XI, 274 pages. 2007.

Vol. 4396: J. García-Vidal, L. Cerdà-Alabern (Eds.), Wireless Systems and Mobility in Next Generation Internet. IX, 271 pages. 2007.

Vol. 4373: K.G. Langendoen, T. Voigt (Eds.), Wireless Sensor Networks. XIII, 358 pages. 2007.

Vol. 4357: L. Buttyán, V.D. Gligor, D. Westhoff (Eds.), Security and Privacy in Ad-Hoc and Sensor Networks. X, 193 pages. 2006.

Vol. 4347: J. López (Ed.), Critical Information Infrastructures Security. X, 286 pages. 2006.

Vol. 4325: J. Cao, I. Stojmenovic, X. Jia, S.K. Das (Eds.), Mobile Ad-hoc and Sensor Networks. XIX, 887 pages. 2006.

Vol. 4320: R. Gotzhein, R. Reed (Eds.), System Analysis and Modeling: Language Profiles. X, 229 pages. 2006.

Vol. 4311: K. Cho, P. Jacquet (Eds.), Technologies for Advanced Heterogeneous Networks II. XI, 253 pages. 2006.

Vol. 4272: P. Havinga, M. Lijding, N. Meratnia, M. Wegdam (Eds.), Smart Sensing and Context. XI, 267 pages. 2006.

Vol. 4269: R. State, S. van der Meer, D. O'Sullivan, T. Pfeifer (Eds.), Large Scale Management of Distributed Systems. XIII, 282 pages. 2006.

Vol. 4268: G. Parr, D. Malone, M. Ó Foghlú (Eds.), Autonomic Principles of IP Operations and Management. XIII, 237 pages. 2006.

Vol. 4267: A. Helmy, B. Jennings, L. Murphy, T. Pfeifer (Eds.), Autonomic Management of Mobile Multimedia Services. XIII, 257 pages. 2006.

Vol. 4240: S.E. Nikoletseas, J.D.P. Rolim (Eds.), Algorithmic Aspects of Wireless Sensor Networks. X, 217 pages. 2006.

Vol. 4238: Y.-T. Kim, M. Takano (Eds.), Management of Convergence Networks and Services. XVIII, 605 pages. 2006.

Vol. 4235: T. Erlebach (Ed.), Combinatorial and Algorithmic Aspects of Networking. VIII, 135 pages. 2006.

Vol. 4217: P. Cuenca, L. Orozco-Barbosa (Eds.), Personal Wireless Communications. XV, 532 pages. 2006.

Vol. 4195: D. Gaiti, G. Pujolle, E.S. Al-Shaer, K.L. Calvert, S. Dobson, G. Leduc, O. Martikainen (Eds.), Autonomic Networking. IX, 316 pages. 2006.

Vol. 4124: H. de Meer, J.P.G. Sterbenz (Eds.), Self-Organizing Systems. XIV, 261 pages. 2006.

Vol. 4104: T. Kunz, S.S. Ravi (Eds.), Ad-Hoc, Mobile, and Wireless Networks. XII, 474 pages. 2006.

Vol. 4074: M. Burmester, A. Yasinsac (Eds.), Secure Mobile Ad-hoc Networks and Sensors. X, 193 pages. 2006.

Vol. 4033: B. Stiller, P. Reichl, B. Tuffin (Eds.), Performability Has its Price. X, 103 pages. 2006.

Vol. 4026: P.B. Gibbons, T. Abdelzaher, J. Aspnes, R. Rao (Eds.), Distributed Computing in Sensor Systems. XIV, 566 pages. 2006.

Vol. 4003: Y. Koucheryavy, J. Harju, V.B. Iversen (Eds.), Next Generation Teletraffic and Wired/Wireless Advanced Networking. XVI, 582 pages. 2006.

Vol. 3996: A. Keller, J.-P. Martin-Flatin (Eds.), Self-Managed Networks, Systems, and Services. X, 185 pages. 2006.

Vol. 3976: F. Boavida, T. Plagemann, B. Stiller, C. Westphal, E. Monteiro (Eds.), NETWORKING 2006. Networking Technologies, Services, and Protocols; Performance of Computer and Communication Networks; Mobile and Wireless Communications Systems. XXVI, 1276 pages. 2006.

Vol. 3970: T. Braun, G. Carle, S. Fahmy, Y. Koucheryavy (Eds.), Wired/Wireless Internet Communications. XIV, 350 pages. 2006.

Vol. 3964: M.Ü. Uyar, A.Y. Duale, M.A. Fecko (Eds.), Testing of Communicating Systems. XI, 373 pages. 2006.

Vol. 3961: I. Chong, K. Kawahara (Eds.), Information Networking. XV, 998 pages. 2006.

Vol. 3912: G.J. Minden, K.L. Calvert, M. Solarski, M. Yamamoto (Eds.), Active Networks. VIII, 217 pages. 2007.

Vol. 3883: M. Cesana, L. Fratta (Eds.), Wireless Systems and Network Architectures in Next Generation Internet. IX, 281 pages. 2006.

Vol. 3868: K. Römer, H. Karl, F. Mattern (Eds.), Wireless Sensor Networks. XI, 342 pages. 2006.

Vol. 3854: I. Stavrakakis, M. Smirnov (Eds.), Autonomic Communication. XIII, 303 pages. 2006.

Vol. 3813: R. Molva, G. Tsudik, D. Westhoff (Eds.), Security and Privacy in Ad-hoc and Sensor Networks. VIII, 219 pages. 2005.

Vol. 3462: R. Boutaba, K.C. Almeroth, R. Puigjaner, S. Shen, J.P. Black (Eds.), NETWORKING 2005. XXX, 1483 pages. 2005.